Biotechnology in Africa

Science Policy Reports

The series Science Policy Reports presents the endorsed results of important studies in basic and applied areas of science and technology. They include, to give just a few examples: panel reports exploring the practical and economic feasibility of a new technology; R & D studies of development opportunities for particular materials, devices or other inventions; reports by responsible bodies on technology standardization in developing branches of industry.

Sponsored typically by large organizations – government agencies, watchdogs, funding bodies, standards institutes, international consortia – the studies selected for Science Policy Reports will disseminate carefully compiled information, detailed data and in-depth analysis to a wide audience. They will bring out implications of scientific discoveries and technologies in societal, cultural, environmental, political and/or commercial contexts and will enable interested parties to take advantage of new opportunities and exploit on-going development processes to the full.

For other titles published in this series, go to
http://www.springer.com/series/8882

Florence Wambugu • Daniel Kamanga
Editors

Biotechnology in Africa

Emergence, Initiatives and Future

Foreword by Dr. Ismail Serageldin

Springer

Editors
Florence Wambugu
Daniel Kamanga
Africa Harvest Biotech Foundation
Nairobi
Kenya

ISSN 2213-1965 ISSN 2213-1973 (electronic)
ISBN 978-3-319-37456-7 ISBN 978-3-319-04001-1 (eBook)
DOI 10.1007/978-3-319-04001-1
Springer Cham Heidelberg New York Dordrecht London

Cover Illustration: www.africabio.com

Printed on acid-free paper

Springer is part of Springer Science+Business Media (www.springer.com)

Foreword

It is now almost 7 years since my colleague, Dr. Calestous Juma, and I authored a report, *Freedom to Innovate: Biotechnology in Africa's Development*, undertaken at the request of African heads of state and government. This main message in the Report of the High-Level African Panel on Modern Biotechnology was a call for regional economic integration to embody the building and accumulation of capacities in order to harness and govern modern biotechnology.

Did Africa heed our call to build the required capacity to harness and apply biotechnologies to improve agricultural productivity, public health, industrial development, economic competitiveness, and environmental sustainability (including biodiversity conservation) in Africa? Have African leaders demonstrated courage and firmness so that their footprints can guide future generations? The answers to these questions are not readily available; however, the book, a collection of papers, edited by my friend, Dr. Florence Wambugu, and her colleague, Mr. Daniel Kamanga, confirm that a lot is happening in Africa's biotech space.

Edited by Africans, with contributions by Africans, the book's unspoken message is that Africa refuses to be left behind. We know that the first century of the new millennium will not only belong to information and communications technology. Biotechnology, and its immense potential to contribute to human and animal health, agriculture and food production, manufacturing and sustainable development, will be an integral part of the arsenal required for Africa's development.

Many of the essays capture the fact that globally, the number of countries that cultivate genetically modified (GM) crops continues to increase. Africa – and more specifically, South Africa, Burkina Faso and Egypt – have joined fast-developing economies such as China, India and Brazil. I am glad to note that since our report, countries such as Kenya and Ghana have passed laws allowing the commercialization of GM crops. Many other countries are undertaking GM research, bringing to almost a dozen, the "African biotech countries".

Africa can boast of a number of GM confined field trials (CFTs) for maize (insect resistance and drought tolerance), cotton (insect resistance and herbicide tolerance), sweet potato (viral and weevil resistance), banana (fungal resistance, bacterial wilt and nutrient enhancement), cowpea (insect resistance), and cassava and sorghum (nutrition enhancement). However, there are still challenges related to strengthening national research systems, especially with regard to infrastructural and human capacity. More critical is the capacity by African countries to decide on how to use these technologies to improve national priority crops.

This book is therefore a clarion call for African leaders from African scientists. While there is no doubt that Africa has turned the corner, there is still need for courage and firmness when it comes to the GM technology. Fortunately, the urgent need to increase agricultural productivity and the increased acceptance of the technology gives African political leaders a window to move to the next level. The cost of not moving forward is too ghastly to contemplate.

Dr. Ismail Serageldin

Acknowledgement

We would like to thank Mr. Rob Garris, Managing Director of the Bellagio Center Programs, for the opportunity accorded to the book contributors to finalize their writing in a serene setting, conducive to focused, goal-oriented work.

Contents

Part VII Political Challenges

Chapter 1
The Importance of Political Will in Contributions of Agricultural Biotechnology Towards Economic Growth, Food and Nutritional Security in Africa

Florence Muringi Wambugu

Abstract Genetically modified (GM) crops or biotech crops, have been commercialized in both developing and industrialized countries since 1996. In 2012, 17.3 million farmers from 28 countries planted crops on 170.3 million hectares (420 million acres) of land, which was 6 % more area, or 10.3 million hectares (25 million acres) more, than in 2011. Breaking new ground, five European countries planted 129,071 ha of Bt maize in 2012. The global value of GM seeds in 2012 was US $15 billion, with commercial grain from biotech crops being valued at about US $150 billion per year. This was described as the fastest growing and adopted technology globally (James C, Global status of commercialized biotech/GM crops: ISAAA brief No. 44. ISAAA, Ithaca, 2012).

In 2012, developing countries—including China, India, Brazil Argentina and South Africa—grew about 52 % of the global biotech crops compared to the industrialized countries' 48 %. Of the 90 % of total farmers (17.3 million farmers), 15 million were smallholders from India and China (a record of 14.4 million in total for both India and China). Only four African countries (South Africa, Burkina Faso, Sudan, and Egypt) have commercialized GM crops, altogether planting only 2.9 million hectares, a very small part of the global or developing countries' average. However, it represents a 26 % increase compared to 2011. Africa's performance in adoption of GM crops for economic benefits is significantly low when compared to other developing countries, hence the need to objectively explore and discuss the underlying issues, while creating awareness on how the current status can be improved. (James C, Global status of commercialized biotech/GM crops: ISAAA brief No. 44. ISAAA, Ithaca, 2012).

Keywords Global food crisis • Agriculture • Africa • Biotechnology

---

F.M. Wambugu (✉)
Africa Harvest Biotech Foundation International (AHBFI), P.O. Box 642, Village Market 00621, Nairobi, Kenya
e-mail: fwambugu@africaharvest.org

F. Wambugu and D. Kamanga (eds.), *Biotechnology in Africa*,
Science Policy Reports 7, DOI 10.1007/978-3-319-04001-1_1,

Abbreviations

AGERI	Agricultural Genetic Engineering Research Institute (Egypt)
AGOA	African Growth and Opportunity Act
Bt	*Bacillus thuringiensis*
CFT	Confined field trials
CGIAR	Consultative Group on International Agricultural Research
EMBRAPA	Brazilian Agricultural Research Corporation
GM	Genetically modified
IP	Intellectual property
NARS	National Agricultural Research Station
NGO	Non-governmental organisation
R&D	Research and development

Global Factors in Favor of Africa's Participation in Agricultural Biotechnology

The ongoing global crisis in food, energy, climatic change, and economic growth has raised the status of Africa as the new frontier of growth and investment for many foreign investors. The global crisis on food deficit in long-term national reserves, and the huge increase in prices of common goods on an annual basis, have opened a new wave in Africa of foreign agricultural investment for local and export markets. While this is being welcomed by African governments, who are facing the challenge of feeding their people, others are concerned about the long-term implications and are describing it as "land grab". The relevance of this to African agriculture indicates an ongoing paradigm shift to viewing agriculture as big business, hence creating the opportunity for biotechnology applications for future growth.

The future growth of agricultural biotechnology in Africa is favored by population growth mainly of young people, most of whom are educated and technologically savvy in terms of knowledge of internet use and so on. Internet navigation skills often equip individuals with access to creditable online resources and debates, which inform and aid individuals to make decisions on biotechnology. Lack of access to and knowledge of such important information has been a prime factor in allowing misinformation, especially from European anti-biotechnology lobby groups. The major foreign investor in African agriculture currently is China, and this trend is likely to continue. Back home, China has invested greatly in agricultural biotechnology. Additionally, China is likely to utilize agricultural biotechnology for their local and foreign investments in Africa, and others are likely to follow suit; overall, Africa will benefit from deliberate technology transfer and from trickle down to the economy.

Future African farmers, whether small-holder or large-scale, are going to move from subsistence agriculture to farming, as a business embracing research-generated technologies of improved seeds and inputs, and are unlikely to discriminate against biotechnology products due to controversies without scientific data to back them up. The current trends in global agricultural biotechnology adoption shows that small-holder farmers (15 million), especially with Bt cotton, have adopted more and benefited more than large scale farmers (2.3 million) (James, 2012). Also, agricultural biotechnology is seed-based and hence more easy to adopt for smallholder African farmers, who might have little education, but have indigenous knowledge and experience on how to handle seed, if other relevant inputs are also available for increased productivity.

Africa and the rest of the world are moving towards a knowledge-based economy, driven by research, fact-based information, data, and expert-based analysis and opinions. This would aid in challenging anti-GM activists, and claims that are not based on scientific data or facts. Agricultural biotechnology is founded on reading, un-coding, copying, and transferring genetic information across. It is intended to benefit humankind, and to be used in medical research for developing products like human insulin to treat diabetes, and in agriculture for developing many genetically modified (GM) crops and products in the global markets. The growing science of genomic analysis, which is being applied to many crop and animal species, is likely to have great influence on many spheres of industry, including agricultural biotechnology.

African agricultural biotechnology adoption has been moderate but strategic in 2012, with four countries in the north (Egypt and Sudan), west (Burkina Faso) and south (South Africa), adopting Bt maize; Bt cotton; Bt cotton; Bt maize, cotton, and soybean; respectively. Many other crops are in research laboratories, in greenhouses, and in confined field trials (CFTs) in various countries, including Kenya (cotton, sorghum, sweet potato, and maize) and Uganda (banana and sweet potato). Future agricultural biotechnology crop opportunities in Africa will continue mainly through the use of Bt genes crop-protection biotechnology, biopesticides, and biofertilizers, and other nutritionally enhanced food crops with vitamin precursors, micronutrients, and protein will also increase. Development of climate-resilient crops will increase especially to address drought, water logging, and salinity. More African indigenous crops which were previously neglected are in the pipeline for biotechnological improvements. These include cowpea, sorghum, cassava and banana. Most of these crops have been developed by African scientists through public/private sector partnerships with multinational biotech companies, using intellectual property (IP) donations, training of African scientists, and technology transfer to African institutions, with funding from African governments, bilateral donor agencies and philanthropic organizations (James, 2012).

In Africa, many farmers of agricultural biotechnology crops, or producers of GM crops are also the main consumers especially for food crops such as Bt maize. This means benefits do accrue to the same people, unlike in western countries where consumers claim they do not benefit from GM crops and products in the supermarkets. Other relevant animal production improvement biotechnology applications include increased milk production hormones, which would mean keeping fewer animals on the limited available land, and reduce environmental degradation associated with overgrazing.

Understanding the Reasons That Cause Delays or Reluctance in African Countries' Decision to Adopt Biotech Crops: The Challenge of Establishing Biosafety Law

Q&A: Is lack of awareness of the benefits of agricultural biotechnology the reason for limited commercialization in Africa when compared to other developing countries?

This book is not about the potential of biotechnology in Africa. However, it is set to explore critically why the "African biotech dream" is being realized so slowly, despite its clear benefits and the great need for this science. This book examines the core issues from an African perspective, and asks key questions such as where are the barriers and bottlenecks, what causes them, and how can they be removed. The purpose is to challenge the obviously accepted and perpetuated theory or hypothesis, and emphasize that the solution is "more awareness creation" on benefits and risks, and letting African countries decide on whether or not to adopt biotech crops. The approach taken is to examine developing countries, including those in Africa, that have successfully adopted biotech crops, and study what they did to overcome the challenges, while trying to understand if the lessons learned can help others who are still struggling.

Q&A: Do African countries believe in agricultural biotechnology, and have they made any investment to confirm their interest?

African countries have made some considerable investment in biotechnology research and development (R&D), mainly in partnership with public and private sector institutions of the north, indicating that they believe in the promise of this technology in impacting agricultural economic growth and development. Where commercialization has taken place, there has been rapid increase in farmer adoption of the technology, and demonstration of impact on benefit to farmers, consumers, and the environment. A brief overview to identify the status of African countries involved in biotechnology R&D indicates that an impressive nine African countries are engaged with biotech crop-improvement at various stages of laboratory, greenhouse, and CTFs. Four of these countries (South Africa, Sudan, Burkina Faso, and Egypt) have commercialized some biotech crops, and have others in the R&D pipeline. The nine countries include Uganda with biotech crop R&D improvements on maize, banana, cassava, cotton, sweet potato, and rice; while Burkina Faso, Malawi, Sudan, and Cameroon are all focusing on cotton improvement. Nigeria is working on biotech crop R&D for improvement of cassava, cowpea, and sorghum; Kenya is conducting biotech R&D for improvements of maize, cotton, cassava, sorghum, pigeon pea, and sweet potato; Egypt has biotech R&D for improvement of maize, cotton, wheat, potato, tomato, sugarcane, rice, and strawberry; and South Africa has biotech R&D for improvement of maize, cotton, cassava, potato, flower bulbs, and sorghum.

Q&A: Has the Cartagena Protocol played a role in the challenges that developing countries including Africa are facing, in the adoption of agricultural biotech crops?

The Cartagena Protocol—which decided that biotechnology poses inherent dangers to biodiversity, and that hence there is a need for a global law that is binding to the government member parties to govern biotechnology—though well-meant by them, due to long-term uncertainties, should have been introduced with clear scientific review guidelines after specific timelines, to determine actions in case such biotechnology dangers as envisaged to the environment did not happen. Apparently because of the lack of such a mechanism over time, now after 17 years (1996–2012) of Biotech crops adoption since signing of the protocol, with no apparent dangers of GM biotechnology to environment, instead of an objective review to ease the stringent regulatory guidelines, more stringent regulatory measures are being prescribed. Stringent Biosafety and regulatory measures were justified at the beginning, when uncertainties existed concerning the long-term food and environmental safety of these foods, but now we have over 17 years of experience (field trails started in 1996), and not a single credible piece of evidence has emerged to prove that GM crops or foods have caused risk to people, animals, or environment; hence the need to ease the burden on the unnecessary costly aspects of the protocol from emerging knowledge. The Cartagena Protocol and the politically charged regulatory and bio-safety law system that goes with it, since it is regulated by parliamentary systems, have been cited as the major barrier to the adoption of agricultural Biotech crops in Africa.

USA refused to ratify the Cartagena Protocol, and is therefore not a party to its political regulatory system. European countries were bound by the Cartagena Protocol but the European Parliament, after considerable delays, decided to engage scientific expert review panels on the safety of these crops. Upon realizing that global adoption was rising every year and they were being left behind, the European Parliament came up with a progressive regulatory system that began to ease the laws on biotech crops. This allowed commercialization in five EU countries in 2012: Portugal (maize), Spain (maize), Czech Republic (maize), Slovakia (maize), and Romania (maize). (James, 2012).

Q&A: Have the European anti-GM lobby groups influenced the adoption of agricultural biotech crops in Africa?

During the first decade of introduction of GM technology that was used to develop biotech crops, the developing countries including Africa were subjected to serious anti-GM lobby campaigns, especially from European NGOs. At times they would recruit and finance some groups, claiming that GM crops were unsafe for human consumption and posed danger to the environment. They also claimed that due to these inherent dangers, the European countries had refused to adopt the technology, and that both the regulators and the regulatory system could not be trusted. Further, they claimed that if any African nation adopted GM crops, it would affect their trade relations with Europe. Furthermore, many of the European donors reinforced their anti-GM position. Although the African governments had started with considerable levels of interest in biotechnology, the confusion, fear, and scare-mongering that followed caused enough disruption to put many of them in a neutral position of wait-and-see, as they did not know whether to adopt the pro-GM stance promoted by the USA or the anti-GM European stance. The USA-generated biotechnology was developed by multinational companies who were easy targets for allegations about creating mistrust, and establishing profit motives. The African politicians and policy makers were also being "pushed" to decide on the highly scientific and technical area of a suitable biosafety law to regulate GM biotechnology that by nature they were poorly prepared for, amidst great promises from one side (USA, Canada, etc.) and big scares from the other (Europe) (Starved for science: How biotechnology is being kept out of Africa/Robert Paarlberg (2008); foreword by Norman Borlaug and Jimmy Carter).

Q&A: How did the developing countries of the East such as Argentina, Brazil, India, and China manage to become major global adopters of biotech crops, while African countries were left behind?

Somewhere in this journey, it seems clear that developing countries such as Argentina, Brazil, India, and China did what African countries failed to do—they took decisive action about using agriculture biotechnology to drive their economic growth. These countries followed this path in spite of the misinformation and confusion generated mainly by European anti-GM lobby groups—many of which were funded by chemical companies and other interested parties that were likely to lose some footing in the market, especially in pesticides, when farmers adopt popular crop-protection biotechnologies such as Bt cotton. These countries managed to break loose from confusion and debate, and took decisive action by putting in place the relevant policy framework to harness the power of biotechnology. However, we must also recall that not all the political systems were able to identify the real issues in these developing countries at the same time; for example, in India and Brazil farmers had to create demand for Bt seeds and policy reforms by illegally purchasing Bt cotton seeds from other farmers in neighboring countries, through the "black market" because of the popularity of the biotechnology. These farmers got GM seeds from their neighboring countries and planted them, and demanded that policy makers become decisive and put the relevant policies in place for them to benefit, instead of them being held back from benefiting, as a result of the indecisiveness of the politicians and policy makers.

Q&A: Why were African countries left behind on the adoption of agricultural biotech crops when compared to developing countries from the East?

The problems in Africa can in a nutshell be traced to the politicizing of the biosafety law about regulating biotech crops, and the challenges that come with that. Progress in Africa can be described as "two steps forward and one step backward". Countries that had taken a lead, such as South Africa, are now taking over 2 years to make a decision regarding permits for applications for field studies of biotech crops. This delays the release of new crops and products, as the process is being politicized as a complaint to the Cartagena Protocol. Apparently seven different government ministers, representing different government ministries (Agriculture, Environment, Education, Mining, Trade, Health, and Tourism and Wildlife) have to agree on a technical matter for the decision to be made. Other countries such as Egypt had in 2010–2011 commercialized Bt maize, but in 2012 were about to reverse that decision due to pressure from European anti-GM lobby groups, when the Arab Spring brought new political leaders and other political changes. The Cabinet of Kenya, at the end of 2012, temporarily put a ban on the import of GM food crops and products, following the publication of a highly controversial scientific article in a European journal (Séralini et al. 2012), claiming that GM foods crops caused cancer. The ban in Kenya was championed by two ministers of the Ministry of Health, both of whom had cancer and were undergoing treatment. The experiments on which the articles were based were soon scientifically challenged by the European Biosafety Agency and many other global scientific institutions, and were found to be compromised and wrong; hence no GM food crop ban occurred in Europe or other countries, except in Africa.

Q&A: Does the political change caused by the elective term of 5 years for Parliamentary members affect biotech crop adoption in Africa?

At the end of an elective term of 5 years in Africa, there is a change of members, and consequently in awareness created, information shared, and knowledge gains invested by individual cabinet ministers. Therefore, any investment in biotechnology crops is lost altogether. This also negatively affects the passing of biosafety bills into laws in Parliament, by causing major delays. This is because cabinet members acting as champions, and their capacity and knowledge of biotech crops, are lost, and with that comes the loss of impact on policy decisions with regard to biotech crop commercialization. The human capacity loss on biotech crops, created by the inevitable political changes, has a profound impact on many aspects of biotechnology policy and biosafety law, including the financial allocation of funds for regulatory personnel and infrastructure, as politicians keep on changing every 4–5 years. This means that where effective expertise and capacity-building have taken place, these changes are disruptive, or they could have a positive effect as someone informed and decisive may come in, although that is rare. Whatever the case, decisions on critical milestones concerning biotech crops and product development, once delayed for long periods, make the process more expensive; and the final product development becomes costly, time-wasting, and discouraging for young scientists in universities and public research institutions, who are hoping to develop some local biotech crops especially for food and nutritional security.

Q&A: What role if any, has the private sector of multinational companies played in African countries, in agricultural biotech crop R&D?

The private sector players have played a major role in biotechnology R&D in and applications in Africa, and have been a source of mixed blessings. The private sector's ability to commercialize biotech crops and products and make them visible, and to deliver positive environmental impact that is well-documented, has helped to keep biotech crop-related work going, despite the anti-GM lobby groups. Nevertheless, it also makes them easy targets for the anti-GM crops lobby groups, making politicians believe there is a profit motive in GM crops, and possibly safety compromises. The fact that currently nearly all GM crops in the market come from the private sector lends itself as a target for critics, without realizing the underlying issues such as heavy costs of product development, emerging from unnecessarily stringent regulations and over-testing that are imposed on these products, to please or satisfy the lobby groups' demands, or precautionary principles, and such controversy are to blame for creating multinational private sector monopoly. The anti-GM lobby groups have indirectly managed to increase the cost of regulatory systems making it difficult for public institutions who mainly work for national good are unable to commercialize their GM research products and subsequently indirectly favoring a monopoly of such products from multinational sector companies who can afford to pay for the costly biosafety regulations. The IP-related issues also favor the private sector companies, enabling them to protect their discoveries and draw value from them, as compared to public institutions. Private sector companies have shown interest in donating or sharing their IP for public good where it does not conflict with their private interest, and many public-good GM crops and products are being developed under these arrangements in many developing countries, including African countries.

Q&A: Is investment in bio-economy through biotech crop R&D, without commensurate investment in biosafety and regulatory policy framework, set to fail?

A visit to African countries' public institutions such as national universities, national agricultural research stations (NARS), related international centers, and even the Consultative Group on International Agricultural Research (CGIAR) centers, will clearly show that there is a high level of investment in biotechnology infrastructure and human capacity, targeting agricultural improvement for national public good. As has been mentioned in some of my other work (see Wambugu, 2001), biotech crop R&D is currently ongoing in nine African countries, targeting crops improvement. Politicians and policy makers in African countries state that they believe in biotechnology to drive agricultural economic growth and development. They have also made considerable investment of national resources in relevant biotech infrastructure and human capacity development, toward training their scientists abroad to ensure they are not left behind in comparison to countries such as South Africa. They want African countries to catch up with other developing countries such as India, Argentina, Brazil, and China that have adopted GM biotechnology and are benefiting greatly from it. Recently, politicians heading parliamentary committees on agriculture and technology in African countries that have not adopted GM crops have also gone on study tours to other developing countries that have commercialized GM technology such as Brazil, South Africa, and

Burkina Faso. Besides the Africa Union through their Agricultural Program called CAADP fully supports Biotechnology, see Freedom to Innovate: Biotechnology in Africa's Development (Juma and Serageldin, 2007). However, these efforts, though exciting for those who participate, have not resulted in any significant impact because of changes that often occur within the parliament, and also because biosafety laws take considerable time to make, during which time those committee members who went for the study tour could have been transferred to other parliamentary committees, before they could bring into effect any beneficial biosafety law development.

Q&A: Is political will the key to the success of biotech crop development through beneficial biosafety policy framework in Africa?

A study of different scenarios in Africa that have led to the limited success of biotechnology development in the commercialization of biotech crops has indicated that political will is essential for any success to occur, because the biosafety policy framework is a political process by design. For example, in South Africa, for a permit to be issued for field testing of GM crop, seven different government ministries, including Agriculture, Health, Trade and Industry, Education, Science and Technology, Labor, and Environment, must unanimously be in agreement. Due to the diversity of ideas that the product being tested raises when they consider biosafety, often the decision made has nothing to actually do with biosafety, but rather with other aspects such as the socioeconomic and political ones, or is simply influenced by anti-GM lobby groups. *Positive political will can be a powerful tool, as shown in the case of Burkina Faso* where the government made a decision to commercialize Bt cotton a few years ago, and now the rest is history, with the country having the fastest rate of adoption of Bt cotton globally. This is an outcome that clearly demonstrates what has gone wrong in other African countries. It is also worth noting that Burkina Faso was assured of a market because of the African Growth and Opportunity Act (AGOA). Some African countries such as Egypt failed to commercialize promising GM products for fear of losing their European markets. Brazil and Burkina Faso are excellent examples of how developing countries can use biotechnology to contribute to sustainable development within the bigger context of agriculture-driven economic growth. *In both cases, political will was necessary, and decisions to engage biotechnology were made at the highest political level in the Parliament.* The decisions included whether to invest in the relevant local infrastructure to gain from international partnerships with multinational private sector companies that had developed the required biotechnologies and held the IP. In the case of Brazil, their top Agricultural Research Corporation (EMBRAPA), was equipped and empowered to participate in biotechnology. The country started like other countries by utilizing multinational private sector GM technologies and negotiating terms on benefit-sharing, resulting in an annual increase in the adoption of GM of up to 20 %, with a total of 36.6 million hectares planted in 2012. Later, Brazil fast-tracked the approval process of six new GM crop products in 2012, including one of their own homegrown GM crops, with a transgenic virus-resistant bean. This shows a very clear strategy at the country level, to engage and benefit from biotechnology, first from international partnerships, while simultaneously developing and strengthening internal country capacities, to support agriculture-driven economic growth, which African countries can emulate. Indeed Burkina Faso has shown strong political will and strategy! Their only challenge has been to their ability to invest in local infrastructure, and capacities to drive their local interests and priority biotechnology needs. In South Africa, there is a strong presence of multinational companies, adequate infrastructure, and human capacities. However, wavering or undecided political systems, as in other African countries, limit the commercializing of GM crops from their local scientific institutions. Egypt is a good example to demonstrate the challenge of a lack of or wavering political will when it comes to biotechnology, and how this can frustrate local scientists and research institutions. Egypt has an excellent biotechnology research institution called the Agricultural Genetic Engineering Research Institute (AGERI), in which the government has made considerable investment towards infrastructure, equipment, and the training of scientists. AGERI should have commercialized several GM crops, both from international partnerships and from their own institution, similar to the Brazilian model. However, due to lack of political will to develop and approve biosafety law through a parliamentary system, very little has happened, and the last attempt by a private sector company to commercialize Bt maize using existing structures was almost reversed after 1 year of commercialization, in 2012. Personal communications with local biotechnology scientists from AGERI indicate that the government is afraid to lose the European market, which is a likelihood if the country embraces GM biotechnology, especially since the anti-GM lobby groups have been exerting a lot of pressure on Egypt, using these threats.

Q&A: What is the way forward for African countries to participate in biotech crop technology in the future?

The only way to resolve a persistent problem is by first diagnosing and understanding the root cause, before attempting corrective or curative measures. What this paper has attempted to do is to clearly show from past experiences that while biotechnology is a scientific, technical process, the biosafety and regulatory policy frameworks are highly political, and that unless and until Africa has the political will and support for biotechnology application, investments made in biotechnology will not be fully realized. African scientists involved in biotechnology need to develop the right strategies and expectations for current and future investments in biotechnology. Understanding and exposing the underlying issues will help African governments make informed decisions when it comes to investment in biotechnology. Hopefully they will also take responsibility with regard to the development of beneficial biosafety and regulatory policy frameworks, to support commercialization of biotech crops, while clearly being aware that it is a politically driven process that can stifle expected impact. It can also help African governments to compare with other developing countries of the East, such as Brazil, India, Argentina, and China, and learn the process that they have followed to be successful in agricultural biotechnology, and take a proactive stance in helping their respective countries benefit from investments made in agricultural biotechnology for food and nutritional security, and also to make economic growth a reality in Africa.

References

James C (2012) Global status of commercialized biotech/GM crops: 2012. ISAAA brief no 44. ISAAA, Ithaca, New York

Juma C, Serageldin IN (2007) Freedom to Innovate: biotechnology in Africa's development: a report of the high-level Africa's development (NEPAD). NEPAD, Addis Ababa/Pretoria

Paarrlberg RL (2008) Starved for science: how biotechnology is being kept out of Africa. Harvard University Press, Cambridge, MA; foreword by Norman Borlaug and Jimmy Carter

Séralini GE, Claira E, Mesnagea R, Gress S, Defargea N, Malatestab M, Hennequinc D, de Vendômoisa JS (2012) Long term toxicity of a roundup herbicide and a roundup-tolerant genetically modified maize. Food Chem Toxicol 50(11):4221–4231

Wambugu FM (2001) Modifying Africa: how biotechnology can benefit the poor and hungry, a case study from Kenya. Nairobi, Kenya

Part I
Commercialized Genetically Modified Crops

Chapter 2
Bt Cotton in Burkina Faso Demonstrates That Political Will Is Key for Biotechnology to Benefit Commercial Agriculture in Africa

H. Traoré, S.A.O. Héma, and K. Traoré

Abstract In 2009, the first year of commercial production of Bt cotton in Burkina Faso, producers planted 129,000 ha, making this the largest introduction of biotechnology on the African continent. The plantation area doubled in 2010 (256,000 ha), but decreased in 2011 (251,580 ha). In 2012, the area of Bt cotton cultivated increased to 300,000 ha. The speeding-up of agricultural biotechnology development in the country is not only due to the political will of authorities, but also because of the determination of stakeholders including scientists, producers, and cotton companies in biotechnology adoption. Therefore, the country's experience provides an excellent example of the processes and procedures which must be gone through for a biotechnology product to be successfully introduced into a developing country where agriculture is a crucial contributor to the gross domestic product (GDP).

The Institute of Environment and Agricultural Research (INERA) and Monsanto conducted controlled experiments with insect-resistant Bt cotton from 2003 to 2006. Success obtained during this seed development program led to evaluating Bt cotton for insect-resistance on a larger scale by commercial farmers in 2007; Bt cotton was commercially released in 2008. Meanwhile, the national rules for safety in biotechnology were adopted in June 2004, and the National Biosafety Agency (NBA) established in 2005. The law on biosafety was passed by the Parliament on March 2006 and promulgated on April 2006. Neighboring countries, especially Benin, Chad, and Mali, would benefit from Burkina Faso's experience, being next in line to introduce Bt cotton.

Keywords Bt cotton • Bollgard II cotton • *Bacillus thuringiensis* • *Helicoverpa armigera* • Biotechnology • Burkina Faso • Cotton value chain

---

H. Traoré (✉) • S.A.O. Héma • K. Traoré
Institut de l'Environnement et de Recherches Agricoles (INERA), 04 BP 8645, Ouagadougou, Burkina Faso
e-mail: hamitraore8@yahoo.com

F. Wambugu and D. Kamanga (eds.), *Biotechnology in Africa*,
Science Policy Reports 7, DOI 10.1007/978-3-319-04001-1_2,

Abbreviations

AAB	African Agency of Biotechnology
AATF	African Agricultural Technology Foundation
ABNE–NEPAD	African Biosafety Network of Expertise – New Partnership for Africa's Development
AIC-B	Inter-professional Cotton Association of Burkina
ANVAR	National Agency for the Valorization of Research Results (Burkina Faso)
ARC	Support to Research on Cotton
ATC	Cotton technical agent
BBA	Burkina Biotech Association
Bt	*Bacillus thuringiensis*
CBD	Convention on Biological Diversity
CC	Cotton correspondent
CCPs	Critical control points
CFDT	Company for the Development of Textile Fibers
CIRAD	International Center in Agricultural Research for Development
CIRDES	International Centre for Research and Development of the Livestock in Sub-humid Zones
CNRST	National Centre for Scientific and Technological Research (Burkina Faso)
CNSB	National Scientific Committee of Biosafety (Burkina Faso)
CORAF/WECARD	West and Central African Council for Agricultural Research and Development
CSIB	Internal Scientific Committee of Biosafety (Burkina Faso)
Dagris	Development of South Agro-Industries
ECOWAS	Economic Community Of West African States
FAO	Food and Agricultural Organization
FARA-SABIMA	Forum for Agricultural Research in Africa – Strengthening Capacity for Safe Biotechnology Management in Sub-Saharan Africa
GDP	Gross domestic product
GMO	Genetically modified organism
GPCs	Groups of cotton producers
IFPRI	International Food Policy Research Institute
INERA	Institut de l'Environnement et de Recherches Agricoles (Institute of Environment and Agricultural Research) (Burkina Faso)
INSD	L'Institut National de la Statistique et de la Démographie
IPS	Industrial Promotion Services
IRCT	Institute for Research on Cotton and exotic Textiles
ISAAA	International Service for the Acquisition of Agri-biotech Applications
LMOs	Living modified organisms

MAHRH	Ministère de l'Agriculture, des Ressources Hydrauliques et de la Pêche
NBA	National Biosafety Agency (Burkina Faso)
NBC	National Biosafety Committee (Burkina Faso)
ONB	National Observatory for Biosafety (Burkina Faso)
PPP	Public/private partnership
RECOAB	West African Network of Communicators in Biotechnology
SOP	Standard operating procedure
SSA	Sub-Saharan Africa
UD	Departmental union
UEMOA	West African Monetary and Economic Union
UNEP–GEF	United Nations Environment Program project–Global Environment Facility
UNPCB	National Union of Cotton Producers (Burkina Faso)

Political Will to Support the Use of Biotechnology

The speeding-up of agricultural biotechnology development in the country is not only the result of political will of authorities, but also the determination of stakeholders including scientists, producers, and cotton companies in biotechnology adoption. Burkina Faso's experience over the past decade provides an excellent example of the processes and procedures required for a biotechnology product to be successfully introduced in a developing country (Vitale et al. 2010); and nearly a decade of coordinated efforts by the various cotton stakeholders was necessary to satisfy a series of technical, legal, and business requirements.

Authorities were engaged at the highest level, as is evident when referring to the speech given by his Excellence Blaise Compaoré, President of Burkina Faso, at the opening ceremony of the Ministerial Conference on Harnessing Science and Technology to Increase Agricultural Productivity in Africa: West African Perspectives. This conference was held in Ouagadougou from June 21 to 24 in 2004, and this speech gives an idea about the political will of the country to encourage the use of biotechnology (Compaoré 2004).

The organization of this meeting was itself a manifestation of the political will of the authorities to move towards biotechnology, and President Compaoré stressed that the exploitation of science and technology to increase agricultural productivity in Africa is pertinent and up to date, for, while agricultural productivity is in increase worldwide, food insecurity still prevails in the African continent. Being the "cradle of humanity", Africa should not miss yet another revolution because, despite the progress made by humanity—thanks to the mastery by human beings of sciences and technology—it is today lagging behind in indispensable discoveries that can ensure the development and well-being of its populations. The Food and Agricultural Organization (FAO) experts' initiative in favor of the use of biotechnologies in the agricultural field to increase productivity and reduce hunger all over the world was relevant. Additionally, the biotechnological revolution should not only be at the origin of recorded progress in the agricultural sector, but also in other domains such as the management of natural resources, health, and industry (Compaoré 2004). The contribution of biotechnology is very important in meeting the future needs of the growing populations in developing countries, particularly in Africa, whose population of 1.1 billion is expected to quadruple by 2,100 to reach 4.2 billion (UN Population Division). To satisfy these population needs and achieve this challenge, the African continent needs to acquire and adapt biotechnologies to the agricultural sector, to increase its current production from 10 to 12 times, and strengthen the collaboration between Burkina Faso's researchers and Monsanto in the experimentation of transgenic cotton (Compaoré 2004). The President also expressed his support for the creation of an African center for research, information, and training in biotechnology, that will strengthen a cooperative relationship between African institutes and institutes in the rest of the world, to give a motivating career to researchers, and attract high-level senior personnel who invest themselves in scientific and technological research.

On June 23, 2004, in his remarks at the same ministerial conference, Dr. J.B. Penn, Under Secretary, Farm and Foreign Agricultural Services U.S. Department of Agriculture, reported an exchange with a journalist: "On the first day of this conference, a journalist here asked me, what is the best thing that this conference can bring to Africa? I answered in one word: Knowledge" (Penn 2004).

For the Secretary, the problem is not the availability of technology in the world that can make a significant positive difference in people's lives here in West Africa, but getting that technology to the people and helping them make use of it—adapting it as necessary and making it easily accessible (Penn 2004). Therefore, this conference provided the opportunity for everyone to share information on technologies, policies, and partnerships, to increase agricultural productivity in West Africa. With regard to biotechnology, Dr. Penn stressed that it is not the American goal to force any technology on anyone. However, there are very positive signs of growing acceptance around the world with regard to products of modern biotechnology and the benefits that they offer. Also, to him, a conference like this, here in West Africa, is a reminder that although friends like the United States may play a helpful role, Africa holds the key to its own development.

The strategic plan for agricultural research in Burkina Faso developed in 1995 [National Centre for Scientific and Technological Research (CNRST) 1995], was carried out in a particular context to boost cotton production. Research programs developed from the demand of producers and existing markets were implemented by combining the most relevant operators. The objectives assigned to the research focused on crop intensification, and improving the profitability of the sector, while ensuring the preservation of natural resources. Thus, since that period, modern biotechnology, including genetically modified (GM) crops, has been identified as a solution to the challenges faced by the cotton sector.

From Experimentation to Commercialization of Bt Cotton in Burkina Faso

INERA in Burkina Faso, as well as the National Agency for the Valorization of Research Results (ANVAR) of the National Centre for Scientific and Technological Research (CNRST), played an important role in the adoption of the Bt cotton technology in Burkina Faso (Zangré 2009; Vitale et al. 2010).

One reason which can explain the interest of Burkina Faso in GMOs is that cotton, the main cash crop of the country was strongly attacked in the 1990s by pests, mainly *Lepidoptera (Helicoverpa armigera)*, which become resistant to insecticides. An important step was also the encounter with Bollgard technology from Monsanto, at the 1999 workshop at Yaoundé. So, ANVAR and the Cotton Program of INERA invited Monsanto to visit Burkina Faso, to present this technology at a meeting of policy makers and people who are interested in cotton in Burkina Faso (Zangré 2009).

From 2000 to 2001, several national workshops for sensitization and awareness creation were organized in Ouagadougou on Monsanto's Bt technology, targeting the key players in cotton (researchers, teachers, ministries, cotton growers, cotton companies, and civil society). Thus, the first major official meeting on Bt cotton was organized in 2000 in Burkina Faso, under the auspices of the CNRST. The workshop was also attended by the concerned ministries (higher education and scientific research, agriculture, animal resources, environment, foreign affairs, finance), civil society, researchers [CNRST, University of Ouagadougou, International Centre for Research and Development of the Livestock in Sub-humid Zones (CIRDES)] Sofitex, the National Union of Cotton Producers of Burkina Faso (UNPCB), and resource persons. This was an opportunity for Monsanto to introduce the new Bt cotton technology that has since been much appreciated by participants, primarily representatives of cotton producers.

Cotton farmers required the acquisition of this new technology and experimentation on it by national researchers in Burkina Faso. However, insofar as Burkina Faso was party to the Convention on Biological Diversity and was about to sign the Cartagena Protocol, the workshop stressed the need for the country to first establish a national biosafety framework as a prerequisite before any importation of GMOs. Monsanto has welcomed this approach, and stated that its strategy is to work only in countries that have regulations on biosafety.

So, when the national rules on biosafety were almost ready, the authorities in Burkina Faso allowed INERA to experiment within the strict rules of confinement, with the Bt cotton of Monsanto known as Bollgard II and Vip of Syngenta, in 2003. Environmental assessments were conducted as part of the input biosafety protocols, along with monitoring the socio-economic impacts of the Bt technology (Vitale et al. 2010). From 2003 to 2005, INERA conducted 3 years of confined field trials (CFTs) to evaluate the biological efficacy of the Bt cotton on the populations of bollworms and particularly those of *Helicoverpa armigera*, and the environmental and health risk within the climate condition specific to Burkina Faso (Vitale et al. 2008, 2010; Traoré et al. 2008; Héma et al. 2009b). For the first 3 years, experiments were carried out with four American varieties (Coker 312 without the Bt gene, Coker 312 with the Bt gene, DP50 without the Bt gene, and DP50 with the Bt gene) at the research stations of Farako-Bâ near Bobo-Dioulasso in western Burkina Faso, and Kouaré, located in eastern Burkina Faso near Fada N'Gourma. The Biosafety Committee, in 2006, approved an additional CFT outside of the INERA research stations, and a series of backcrosses showed a successful transfer of the Bt gene from the American varieties to the three improved Burkinabé cotton varieties of INERA (FK 37, FK 290, STAM 59 A).

In 2006, for the fourth year of experimentation, Saria research station in the central zone, and Boni, a seed farm located 120 km from Bobo-Dioulasso on the axis of Bobo-Dioulasso-Ouaga, also hosted trials, in addition to the two research stations mentioned above. All the experimentation sites are located between the isohyets 800 mm and 1,000 mm.

The results of the bioassay obtained from 2003 to 2006 showed that the presence of the Bt gene in the American varieties and landraces helped to significantly reduce infestations of *Helicoverpa armigera*, *Diparopsis watersi*, and *Earias* spp. at Farako-Bâ, Kouaré, and Saria (Institute of Environment and Agricultural Research (INERA) 2007; Vitale et al. 2008, 2010; Traoré et al. 2008; Héma et al. 2009b). For the control of defoliator populations of *Syllepte derogata*, *Spodoptera littoralis*, and *Anomis flava* on the three sites, the efficacy of the Bt gene was equivalent to that of the standard pest control regimen of six sprays, and the cotton containing the Bt gene had no effect on the group of piercing and sucking Insects. A complementary program based on the last two insecticide treatments applied against the piercing-sucking pests of Bt cotton was as effective as the conventional six treatments.

With regard to economical evaluation, Bt cotton saves the first four treatments, which reduces the cost of insecticide protection of cotton by 67 % (INERA 2007). Bollgard II cotton provided a significant yield advantage of 14.7 % over conventional cotton, and the Bollgard II cotton had a significantly higher profitability than conventional cotton (Vitale et al. 2008, 2010; Traoré et al. 2008). The average profit obtained was 33,000 French CFA/ha compared to the insecticide protection program of the producers (six insecticide treatments per hectare).

A study was carried out on the impact of Bt cotton on the environment, auxiliary fauna, gene flow, and the effect of toxins produced by Bt cotton on populations of honey bees (INERA 2007). It reveals that the auxiliary fauna is not influenced by the presence of Bt cotton, and at 15 m, one cannot find more than 0.5 % of pollen from transgenic cotton.

With regard to the activity of honey bees, no significant abnormality or disturbance was noted in the behavior and the pace of development of bee colonies on Bt cotton plots, compared with the bees of conventional cotton fields (INERA 2007). On the contrary, a positive trend seems to be emerging for the mitigation of aggression of colonies, and for the increase in amount of honey stored, and speed of beam-capping. The comparison of lipid content between samples of cotton varieties DP50 and Bollgard II shows that there was no significant difference, but in contrast, the two varieties (DP50 and Bollgard II) have significantly different levels of protein. The study of the acute toxicity of endotoxin in rats reveals that the samples were of low toxicity: LD50 limit oral > 3,000 mg/kg.

A key stage during this seed development program was the transition from conducting highly controlled research trials to evaluating Bt cotton for insect-resistance on a larger scale by commercial farmers (Traoré and Héma 2011). All agronomic advancements conducted by INERA need pre-extension testing before release, to verify real benefits for commercial farmers. In 2007, after the Bt cotton had successfully completed CFTs at research stations, 20 farmers (10 in the Sofitex zone, 6 in Socoma zone and 4 in Faso Cotton zone) from across Burkina Faso, enrolled in pre-extension tests in CFT conditions, for demonstration to farmers. An average yield increase of 20 % was obtained in 2007, and the NBA in June 2008 authorized the commercial planting of Bt cotton in Burkina Faso (Vitale et al. 2010), marking the first commercial use of Bt cotton in the country, and the third commercial release of a bioengineered crop in Africa. So, in the 2008 cotton-growing season, Sofitex and its contract seed producers planted 15,000 ha of the two local varieties containing the Bt gene to produce seeds for the next year. The way was then paved for the 2009 commercial planting of 125,000 ha of Bt cotton in Burkina Faso, the most extensive single-year biotechnology launch in Sub-Saharan Africa (SSA) to date (Vitale et al. 2010).

Bt cotton was commercialized in 2009, and a license agreement for 3 years renewable for production and distribution of Bollgard II seeds was signed between Sofitex and Monsanto. A collaborative 2-year agreement renewable for accompanying Bollgard II technology was signed between Monsanto and INERA, covering various areas (production of breeder seeds, defining technical itineraries adapted to the cultivation of Bollgard II, defining technical itineraries suited to the production of Bollgard II seeds, monitoring the efficacy of Bollgard II in field conditions, monitoring pests susceptibility to Bt toxins, monitoring non-targeted organisms by toxins, setting suitable refuge zones, and training and information of advice-support staffs and producers).

Evolution of Bt Cotton Hectarage, Seed Supply System, and Cohabitation with Other Cotton Crops

Evolution of Bt Cotton Hectarage

After successfully testing Bt cotton varieties in research stations from 2003 to 2006, in 2007 the demonstration phase of Bt cotton throughout the country, with 20 farmers considered as leaders, was an occasion to organize field days to familiarize producers with this new technology (Traoré and Héma 2011). Farmers were very enthusiastic about this experience; therefore, in 2008, the seed production of two transgenic local varieties was conducted on 8,500 ha. After various tests, chemical processing and packaging, the seeds produced in 2008 were used in 2009 to sow 129,000 ha, or 31 % of the total area under cotton (Table 2.1), of what was the first commercial production (Sofitex 2012). In 2010, the total area under cotton had decreased, but the area under Bt cotton doubled from 129,000 ha in 2009 to 256,000 ha in 2010; that is, 66 % of the total area. In 2011, Burkina Faso experienced an unprecedented crisis during the period of implementation of crops, despite the improvement seen in the global market of the fiber. Some farmer organizations were claiming lower sale prices for inputs and higher purchase prices of seed cotton. Therefore, they refused to sow and attacked those who did not respect their boycott. This situation, which was settled thereafter, had a negative impact on the distribution of Bt cotton seed. Thus, despite the increase in the total area of cotton, there was a slight decrease of about 5,000 ha in the area of cultivation of transgenic cotton. The amount of seeds produced in 2011 was not enough to satisfy all requests for the year 2012, as a result of which Burkina Faso's cotton growers planted 300,000 ha of Bt cotton, which represents 57 % of the total area under cotton.

Seeds Supply System

The seeds supply system developed in the context of conventional cotton production is the same for the production of transgenic Bt cotton (Sofitex 2012). Indeed, INERA annually produces at least 500 kg of breeder seeds that are available to Sofitex for producing foundation seeds at the Boni seed farm. The foundation seeds are then given to seed producers recognized for their compliance with technical innovations and recommendations for cotton production. They receive an additional premium for compliance with all the good agricultural practices applied to the seed. Seed producers then produce seed cotton which will provide certified seeds after a series of tests for the presence of genes of interest and consistent germination. Seed treatment and packaging are done by Sofitex, which supplies all producers according to their specific needs.

Cohabitation of Bt Cotton with Other Cotton Crops

It is also interesting to see the cohabitation of the three types of cotton crops. In Burkina Faso, three types of cotton co-exist in the same ecosystems: conventional cotton, consisting of conventional local varieties, which uses conventional inputs such as mineral fertilizers and pesticides (herbicides, insecticides, fungicides); the transgenic Bt cotton, which consists of landraces back-crossed with the Cry gene from *Bacillus thuringiensis*, and also uses conventional inputs such as mineral fertilizers and pesticides; and organic cotton made of conventional varieties,

Table 2.1 Evolution of the Bt cotton area from 2009 to 2012

Year	Total area under cotton (hectares)	Area under Bt cotton (hectares)	Rate of Bt (% of total)
2009	420,000	129,000	31
2010	386,000	256,000	66
2011	429,000	251,580	59
2012	530,000	300,000	57

which does not use any synthetic chemicals without the label "organic". This type of cotton is quite minimal (less than 1 % of the total area of cotton in Burkina Faso), and uses organic fertilizers and organic pesticides. It is mainly grown by women's groups which do not have access to enough input credit for the production of the other two types of cotton. Cotton producers are free to choose the type of cotton they wish to grow. Only in the case where Bt cotton is chosen, it is necessary to install approximately 20 % of conventional cotton, treated according to the conventional standard program recommended, that is six insecticide treatments starting 30 days after emergence, with an interval of 15 days between treatments. This measure, called 'refuge area', allows the dilution of resistance genes to Bt toxins, ensuring the sustainability of the technology. An isolation distance between the different types of cotton grown is respected to avoid pollution. Also, crops are separated and transported separately to the ginneries. Cohabitation between cotton crops generates costs that producers want to minimize, and their establishment near to each other is avoided. Research is underway in INERA to determine crops and their percentages in terms of areas to even replace conventional cotton for the sustainability of Bt technology.

Legislative Framework Implementation by the Bodies in Charge of Biosafety Law

Burkina Faso has been part of many international commitments, and participated in their implementation. In 1992 in Algiers, Burkina Faso co-founded with 15 other member states the African Agency of Biotechnology (AAB). The AAB works to strengthen the capacity of member countries in biotechnology and for the promotion of commercial biotechnology. Ever since, the country has manifested its intention to promote the development of biotechnology (Zangré 2009).

The Cartagena Protocol on Biosafety was derived from the Convention on Biological Diversity (CBD) adopted in May 1992 in Nairobi (Kenya), and Burkina Faso signed the CBD in 1993.

In the application of the precautionary principle on environment, as stated in the Rio Declaration of 1992, the establishment of an international instrument in the form of a protocol, which was inclusive of permission to manage biosafety issues, including the cross-border movements of GMOs, became imperative.

Thus, in 1995, the Convention designated a group of 15 experts, respecting regional balances, to address the issue. Burkina Faso and South Africa represented South Saharan Africa in Cairo (Egypt) in the consultation that recognized the novelty of living modified organisms (LMOs) (later called GMOs), and recognized the need to negotiate an international protocol to manage biosafety issues that would come from biotechnology. The country has participated in the consultation of the United Nation Program for Environment, with regard to the adoption of transitional guidance, pending after the Protocol. From 1996 to 2000, from Aarhus (Denmark) to Montréal (Canada), Burkina Faso has contributed to the non-limited group in terms of composition, on the Cartagena Protocol. The country was also part of the 2001 panel of experts for the development of the Model Law on Safety in Biotechnology of the African Union.

Burkina Faso has participated from the beginning to the end of this long negotiation process involving several meetings, through the CNRST, the Ministry of Environment and Life Framework. The negotiations led to the adoption of a binding protocol in 2000 in Nairobi, which came into force on September 11, 2003. The Government of Burkina Faso ratified the Cartagena Protocol on August 04, 2003. In the application of the Protocol, the signatory countries undertook to implement national biosafety frameworks whose scope was not below the protocol. In 2004, Burkina Faso ratified the legal instruments of the AAB, and recently adopted, in 2010, the Supplementary Protocol on Liability and Redress, known as the Nagoya–Kuala Lumpur Protocol.

A workshop held on March 2000 set up a temporary committee chaired by ANVAR, CNRST, and was composed of one representative from the Ministry of Environment and Life Framework, University of Ouagadougou, INERA, and a representative of civil society as a resource person, to think and develop a national biosafety framework. This committee worked for 2 years, until 2002, to provide a document entitled "National rules on safety in biotechnology", which was enriched by the recommendations of the United Nations Environment Program project–Global Environment Facility (UNEP–GEF), carried out by the Ministry of Environment and Life Framework for the implementation of the national biosafety framework. Subsequently, the rules were validated by a national workshop held in Ouagadougou in November 2003.

A National Framework on Biosafety, the result of extensive national consultations with stakeholders and all categories of users of GMOs and derived products (ministries, civil society organizations, NGOs, traders) was created in 2003.

The adoption of National Rules for Safety in Biotechnology on June 18, 2004 by decree by the Government of Burkina Faso represents a significant step forward in the regulation of GMOs in the country. A National Biosafety Committee (NBC) was implemented in 2004. Inspired by national rules, a law on security in Biotechnology in Burkina Faso was passed by the National Assembly in March 17, 2006 and promulgated on April 13, 2006 with eight titles and 75 articles. The biosafety law has been translated into three local languages. The implementation of the NBC is done through different regulation bodies: National Biosafety Agency (NBA)—created in 2005, National Scientific Committee of Biosafety (CNSB), National

Observatory for Biosafety (ONB), and Internal Scientific Committee of Biosafety (CSIB).

Overview of the Agriculture and Cotton Sector in Burkina Faso

History of Cotton Production in Burkina Faso: Independence to the Present

Burkina Faso is a landlocked West African country with an area of 274,000 km^2 (of which 9 million hectares is arable land), a population estimated in 2006 at 14,017,262, and an annual growth rate of 3.1 % [L'Institut National de la Statistique et de la Démographie (INSD) 2008]. The rural sector is important in the national economy, since 86 % of people are farmers. Agriculture contributes 40 % to the GDP, with 25 % for crop production, 12 % for livestock, and 3 % for forestry and fishery (Ministère de l'Agriculture, des Ressources Hydrauliques et de la Pêche [MAHRH] 2008). Agriculture contributes to 44.7 % of total household income, with 24.3 % from crop production and 20.4 % from livestock. Cotton is the principal cash crop in Burkina Faso, generating over US $300 million in annual revenues. It accounts for between 5 % and 10 % of the GDP in Burkina Faso (International Food Policy Research Institute [IFPRI] 2006), and represents more than 50 % of the country's export earnings (INERA 2002).

Cotton has been grown for more than a century in the Sahelian and Sudanian savannas of West Africa (Club du Sahel 2005). Cotton has played an important role in the economic development of many countries in West Africa, and it still remains an important source of income for many of them. Grown for its fiber and the oil extracted from the seed, cotton is the main export crop in many West African countries. In West Africa, cotton cropping is the main economic activity for more than 1 million households, and sustains some 10 million farmers. The bulk of the production is carried out by small farmers producing their cotton under rain-fed conditions on areas of 1–2 ha and generally practicing cotton–cereal rotation. Traoré et al. (2008) report that cotton has been the primary catalyst to economic development because where it is grown, rural infrastructural growth has been seen. Therefore, cotton has been the driving force behind the construction of roads, schools, banks, and hospitals in rural areas. Africa has hundreds of varieties of cotton, some of which date back to the tenth and thirteenth centuries. Grown in rain-fed conditions on about 2.4 million hectares, cotton production in West Francophone Africa has been for nearly 40 years the main engine of economic growth (Chetaille 2006). With a production of 730,000 t of seed cotton in 2005, Burkina Faso is now the first cotton producer in Africa, and the development of this crop has been a success, which has helped to reduce poverty in areas where it is practiced.

The Withdrawal of Government, and the Organization of the Value Chain

Cotton production on a large scale began in the 1950s with the French Company for the Development of Textile Fibers (CFDT), now called Development of South Agro-Industries (Dagris), which introduced new cotton varieties (American Upland cotton) for the purpose of the textile industry. The association of Upper Volta-CFDT was created, and lasted from 1970 to 1979. On June 20, 1979, the Government of Upper Volta created the Society of Textile Fibers, which in 1984, became Sofitex, the Burkinabè Company of Textile Fibers. Dagris performed nearly all activities including production of seed cotton and stabilization of prices and incomes, with the exception of agricultural research. From the "one-stop" cotton farming system in which Sofitex provided all the production inputs and also purchased all the seed cotton from the farmers, the Government of Burkina Faso divested itself in 2002 of complete control of the cotton sector (Traoré et al. 2008). In late 2004, Sofitex sold the production area of the Centre to the consortium of Industrial Promotion Services (IPS) and Paul Reinhart AG, and the East Zone to Dagris; Sofitex has, meanwhile, maintained its role in the West Zone. It should be noted that the institutional aspect, however, has played a major role in the geographical distribution of seed cotton production. Cotton production always begins in areas where the government guaranteed to producers, through the intermediary semi-public cotton company, the purchase of any seed cotton produced, and thus created some income security appreciated by producers.

Organization of Cotton Production

Production Cycle

The production of seed cotton involves producers, the extension services of cotton companies, and agricultural research. The success of this production requires the organization of the value chain and the various actors (cotton companies, producers, cotton research, carriers, and so on) for the supply of inputs, and marketing. Apart from cotton seed, other inputs such as fertilizers, cotton insecticides, herbicides, and treatment devices are subject to import through international tenders.

The breeder seeds are produced by the cotton research program of INERA and multiplied into foundation seeds at the farm-level by cotton companies (Sofitex). Certified seed production is provided by individual producers in informal contracts with the cotton companies. The establishment of seeds and insecticides takes into account soil and climatic conditions, and the parasitic infestation level of each production area. Producers play an important role in the production cycle. Cotton is grown in a strictly rain-fed regime by smallholders practicing mainly animal traction. Over 90 % of farms have less than 5 ha, and the average area covered by the cotton is between 1 ha and 2.5 ha per farm. One can easily distinguish between collective cotton farms and individual cotton farms. Collective farms include individual producers organized into groups of cotton producers (GPC), with each group containing 15–40 or more farmers. Individual farm producers are not members of GPCs. There are approximately 325,000 cotton farmers [National Union of Cotton Producers Institute of Environment and Agricultural Research INERA (2007)] in Burkina Faso.

The Institute of Environment and Agricultural Research INERA is a group of farmers from the village to the province (Traoré 2007). According to Mr. François B. Traoré, former President of UNPCB, “the group began in 1996 and in 1998 we established the office of the National Union”. In 1999, the producers became a shareholder in Sofitex, the national cotton company, and in 2005 they became shareholders in the new companies that were put in place after the privatization. So the cotton sector is jointly owned by the Burkina Faso Government, the private sector, producers, and three companies that are operating in three different zones, each maintaining the “one-stop” cotton farming system (Traoré et al. 2008). The production of Burkina Faso was around 116,000 t of seed cotton in 1996, but in 2005–2006 production in the country reached 713,000 t. Thus, Burkina Faso became the first African country to be a producer of cotton. The cotton price is negotiated among the principal stakeholders, giving producers a significant voice in determining cotton price levels, and that has created a climate of trust between the producers and the companies, which is very important in terms of trade.

The research also plays an important role in the sector. The CFDT was created in 1949, began its activities in Upper Volta in 1951, and then benefited from the collaboration of the French Institute for Research on Cotton and exotic Textiles (IRCT) of the International Center in Agricultural Research for Development (CIRAD), established in 1946. The objectives of CIRAD were to study and breed for high-yielding varieties adapted to different regions. Cotton research in Burkina Faso was led mainly by the IRCT until the year 1985.

The cotton research is actually implemented by the cotton program of INERA, which reports to the Crop Production Department, whose mission is to develop cotton varieties with good productivity in the field, and with technological characteristics that meet the requirements of international market. This program is under the responsibility of a Program Manager, supported by section leaders. Major research activities are conducted in these four sections: varietal improvement, agronomy and cultural techniques, crop-protection, and agro-socio-economics.

There is a mechanism of programming, implementation, and dissemination of research findings. Programming research activities is done for a period of 3 years. These activities are part of the strategic plan of the national research, and take into account the concerns of the sector.

In 1988, Sofitex and INERA signed an agreement entitled “Support to Cotton Research [Agricultural Research Corporation (ARC)]”. With the signing of the contract plan between Sofitex and the Government in 1993, a new way of funding cotton research has been proposed on the basis of 1.5 F CFA/kg[1] of fiber produced. Although this ratio has never been reached, the cotton research has consistently received funding from the cotton sector.

The privatization of the cotton sector in 2004 has not fundamentally changed this mechanism of funding research. The Inter-professional Cotton Association of Burkina (AIC-B), which includes the three cotton companies (Sofitex, Socoma, and Faso Cotton), and the UNPCB decided to finance cotton research that presents its results each year and submits its work plan and budget to the Management Committee of the AIC-B, which in turn conducts an appraisal.

Research activities on cotton are funded by the value chain through the Protocol "Support to Cotton Research (ARC)" signed between the management committees of AIC-B and INERA, and their activities take into account cotton-based production systems.

Evolution of the Extension System

Extension has been provided since 1992, primarily by cotton companies that have services in the field. The current system consists of approximately 120 cotton correspondents (CC) and 300 cotton technical agents (ATC). There is about one CC by departmental union (UD) and one ATC for about 50 groups of cotton producers. Programming the extension activities is done annually and it covers the technical itinerary and the use of new technologies. Extension is more concentrated on cotton. The Ministry of Agriculture provides some extension services through an advice–support to producers for all farm activities and on issues not specific to cotton. The extension system ensures a continuous advice–support. This form of extension seems only mildly effective for certain categories of producers, given the level of professionalization of some of them. Thus, the agricultural farming council is an alternative for a more appropriate extension.

Major Achievements in Cotton Research

Among the main achievements, one can notice the presence of a gene bank of over 200 varieties (with gland or glandless), and ten varieties of colored fibers created at the Farako-Bâ research station (INERA 2007). These colored varieties are not cultivated, but were created to anticipate the demand of the international market.

In collaboration with Monsanto, three caterpillar-resistant varieties were newly created, and their experimentation is ongoing.

---

[1] F CFA states for: Franc of the African Financial Community, a currency in 8 West Africa francophone countries (Benin, Burkina Faso, Côte d'Ivoire, Guinea Bissau, Mali, Niger, Sénégal, Togo).

Three cotton varieties are grown in Burkina Faso: two Burkinabè varieties (FK290 and FK37) are grown in areas where annual rainfall exceeds 800 mm, and a Togolese origin variety (STAM 59A) is grown in areas of low annual rainfall (INERA 2007).

Research in agronomy has led to the optimization of mineral fertilizer formulas in cotton cultivation. Problems related to the fertilization of cotton are studied to propose formulas for mineral fertilizers and organic manures in suitable farming systems based on cotton and cereals. This work has led to propose "bulk blending" of cotton fertilizer, which has a similar efficacy to that of complex fertilizer which was previously used exclusively.

Competition from weeds due to delays in weeding resulted in yield losses of about 200 kg/ha of seed cotton per decade, along with an impairment of cotton quality (INERA 2007). More than 20 new herbicide formulations were popularized, and the use of herbicides for 3–4 consecutive years results in reduction in weeding time from 20 h/ha to 40 h/ha on plots with medium to high weed infestation, and a yield improvement of more than 11 %.

The main achievements in entomology take into account good knowledge of periods and durations of proliferation of population of the main insect pest species (INERA 2007). For bollworms, the period of abundance of species is during the first 3 weeks for *Diparopsis*, and the last 5 weeks for the *Helicoverpa* and *Earias*, while for the defoliators, the period of abundance is the entire cotton cycle. With regard to piercing and sucking insects, aphids and whiteflies are present throughout the cycle, with a remarkable abundance at the beginning and end of the cycle for the first, and end of cycle for the latter.

Studies undertaken on the social and economic importance of cotton showed that cotton represents 61–65 % of the income of producers, but only 31 % of this income is reinvested in agricultural activities (INERA 2007). The rest is divided between real estate and luxury, social spending, and general purchases. Cotton is also a relay to the development of traditional cereals. Studies have shown that income management is a key factor to ensure the sustainability of the farm. This requires the development of a suitable farm council, which will enhance the ability of the producer to make the diagnosis of its farm in order to consider actions to improve incomes. The impact studies undertaken have shown that investment in agricultural research have been very profitable. For an investment of 10.19 billion CFA francs over 20 years (1980–2000) in research and extension on cotton, profits generated for producers and consumers are estimated at 59.606 billion F CFA. Therefore, investment in cotton research and extension is a viable use of public funds and development assistance.

Constraints to Cotton Production and Challenges

Despite the great contribution of cotton (conventional cotton for many years and recently the Bt) to the agricultural sector of Burkina Faso, the country is beset by a number of challenges that undermine production, including vulnerability to climate shocks, low yields due to the extensive nature of farming practices, drought, poor soil, weeds, the low level of technology transfer, the difficulties faced by producers to access new technologies, and lack of infrastructure and inadequate credit.

In addition, the cotton plant faces serious damage by many pests, particularly insects feeding upon the leaves and fruits, and yield losses on most cultivated varieties may represent 90 % of potential yields in conditions where no control measures against pests and diseases are undertaken (Michel et al. 2000). In Burkina Faso, most damage is due to two principal groups of caterpillar (*Lepidoptera*) pests which can be distinguished by their feeding preferences: (1) bollworms or fruit-feeders, the most prevalent of which are *Helicoverpa armigera* (*Hübner*) (old world bollworm), *Diparopsis* spp. (red bollworms), and *Earias* spp. (spiny bollworms), and (2) defoliators, which are primarily *Syllepte derogata* (*Fabricius*) (cotton leafroller), *Anomis flava* (*Fabricius*) (looper), and *Spodoptera littoralis* (*Boisduval*) (cotton leafworm) (Héma 2004).

From 1980 to 1995, cotton in the West African sub-region including Burkina Faso was protected from caterpillar damage by the application of binary insecticides containing both pyrethroids and organophosphates, often associated in the same treatment (Martin et al. 2000; Héma 2004). But since 1995, applications of insecticides have failed to control insects, particularly the larvae of *H. armigera*, confirming the presence of resistance (Martin et al. 2000; Héma 2004) and leading to an increase in the number of insecticide applications (from 6 to 8) by cotton growers, to reduce pest infestation. The consequences were higher production costs and adverse effects on human health and environment. Vitale et al. (2006) reported that, in a typical year, the Burkina Faso cotton sector uses over US $60 million of chemical-based pest control products. Despite that, a recently conducted study in Burkina Faso found significant pest damage on fields that were protected using a standard regimen of six seasonal sprays. Thus, the introduction of GM cotton plants expressing the Cry1Ac and Cry2Ab toxins (Perlak et al. 2001; Greenplate et al. 2003; Héma et al. 2009a), which have a different mode of action as opposed to that of pyrethroids, became an interesting and effective alternative to chemical control. Different studies have shown in other areas of the world that transgenic cotton or transgenic maize use has greatly reduced pesticide treatment, while effectively controlling insect pests [Pray et al. 2002; International Service for the Acquisition of Agri-biotech Applications (ISAAA) 2010]. So, to reduce losses due to pest damage, authorities and stakeholders in the cotton sector including scientists, producers, and cotton companies of Burkina Faso decided to evaluate the Bt technology. The results from the evaluation led to the adoption of Bt cotton.

The Way Forward: Lessons Learned from Burkina Faso's Experience

With regard to biotech crops, today the African continent represents by far the biggest challenge in terms of adoption and acceptance (James 2010), and the decision in 2008 by Burkina Faso to grow 8,500 ha of Bt cotton for seed multiplication and initial commercialization (Vitale et al. 2008, 2010; Traoré et al. 2008; Traoré and Héma 2011), and for Egypt to commercialize 700 ha of Bt yellow maize hybrid for the first time (Sawahel 2008; ISAAA 2007, 2010; ABNE 2010) was of strategic importance for the African continent. Burkina Faso and Egypt took the leadership in West Africa and North Africa respectively, in addition to South Africa for Southern and Eastern Africa, for commercializing biotech crops in 2008. This broad geographical coverage in Africa is of strategic importance in that it allows the three countries to become role models in their respective regions, and allows more African farmers to become practitioners of biotech crops and to be able to benefit directly from "learning by doing", which has proven to be such an important feature in the success of Bt cotton in China and India (James 2010).

In Egypt, during a field visit organized by the Egypt Biotechnology Information Center which witnessed a gathering of 50 scientists, maize breeders, and private company representatives, as well as 100 farmers in Bt maize fields in the Sharkia Delta, Egypt last August 23, 2010 (ISAAA 2010), Prof. Magdy Massoud from the Alexandria University explained to the audience that the Bt maize variety could be planted at any time of the season, as it is resistant to the maize borer. He also added that the new variety increases corn yield by up to 30 %. In response, farmers had expressed their interest in the variety, noting that they could use less pesticide and labor, and have higher yields in addition to the benefit of it being an environmentally friendly crop. Moreover, after intensive Bt maize field trial studies in over 36 maize-growing areas in Egypt, Dr. Magdy Abdel Zaher pointed out that "The use of biotech maize saves on pesticide usage, gives almost 100 % protection from stem borers and increases yield by 30–40 % over the conventional maize varieties" (Abdallah 2010). The future of biotechnology seemed brighter for the benefit of farmers not only in the context of Bt maize, but also for potato, cucurbits, wheat, rice, and tomato, which are in the pipeline (Abdallah 2010).

In Burkina Faso, to know how the Bt cotton technology adoption complied with producers' will, their point of view must also be understood. In response to the question "Could transgenic cotton be a solution?" Mr François B. Traoré, former UNPCB President, in 2007 replied, "Yes, Bt cotton can be one of the solutions. You know the Bt cotton was developed by researchers who are at the same time traders. Their goal is to reduce the use of pesticides for the treatment of parasites in our fields. This research is accompanied by action plans that will allow these companies to make a profit. Meanwhile, manufacturers and sellers of pesticides that are reducing their future earnings cannot be happy. The second aspect is the novelty. All that is new is scary and raises questions. There are people that evoke our dependence vis-à-vis those firms producing Bt cotton seeds. But listen, it's trade, both parties must benefit from this trade. If this is not the case, we'll see. We are not associated with these companies for eternity. If we do not take advantage of this trade, we will stop. There are also people who evoke our dependence vis-à-vis these firms with regard to seed production. Do you think today that the producer who uses the seeds he obtained from his grandfather may have good yields? This simply means that the improvement of cotton seeds by GMOs can be a way out for us. This will eliminate the cost benefits associated with pesticides, reduce diseases related to the use of these pesticides, we save time and many other things. The issue of dependency is not real. I'm just saying that money does change hands. It is the pesticides seller who will lose, to the benefit of the Bt cotton seeds seller. Realistically, there are people who wonder if you can eat GMO food. The American are richer than us, the Chinese are richer than we, Hindus are richer than us and yet GMOs are cultivated and consumed. So why do we think it will lead to death? In view of all these aspects, I think we have nothing to lose by adopting GMOs. Rather, it can improve our outcome".

Nevertheless, the Egyptian experience was unfortunately stopped in 2010 because of the non-promulgation of any law, even though a drafted national biosafety law has existed since 2004 (Sarant 2012), revealing the complexity of processes and procedures necessary to support the adoption of an agricultural biotechnology product.

Innumerable lessons can be learned from these two countries' experiences, and can help others to go forward in the adoption and deployment of biotechnology in Africa.

Burkina Faso underwent a unique experience that can show that biotechnology can overcome challenges in legal frameworks, technocratic bureaucracy, and can be supported and sustained by business models that link the private sector to small-and medium-sized producers in developing countries. The surrounding countries such as Mali, Togo, Benin, and Ghana would probably benefit as much as Burkina Faso in Bt cotton technology and could be next in line in the introduction of the technology, once legal frameworks are established. The adoption of biotechnology was facilitated by not only the political will of the authorities of the country, but also by the efforts of all the stakeholders. Indeed, the country has benefited from the support of many players in moving biosafety forward. Among those inside and outside the country are the government, lawyers, universities, researchers, NGOs, national and international activists, the African Biosafety Network of Expertise—New Partnership for Africa's Development (ABNE–NEPAD), the West African Monetary and Economic Union (UEMOA), the Forum for Agricultural Research in Africa—Strengthening Capacity for Safe Biotechnology Management in Sub-Saharan Africa (FARA-SABIMA), West and Central African Council for Agricultural Research and Development (CORAF/WECARD), the Economic Community Of West African States (ECOWAS), African Agricultural Technology Foundation (AATF), and the International Service for the Acquisition of Agri-biotech Applications (ISAAA).

In the development process of Bt cotton, sensitization and communication were done by scientists, with great contributions from the Burkina Biotech Association

(BBA), and the West African Network of Communicators in Biotechnology (RECOAB). These two associations, with the financial support of ISAAA, were involved in advising and training policy and decision makers (Members of Parliaments), and journalists of Burkina Faso and other African countries. The President of the BBA, Professor Alassane Séré, was right in arguing that "If Africa missed the first green revolution, it should not miss that concerning the contribution of biotechnology to the development of agriculture, and we have to ask ourselves: Will Africa still run behind a new green revolution? (Séré 2007)"

The Bt genes have been transferred into the local landraces (FK 37, FK 290, STAM 59A), and adapted to the agro-ecological conditions of the country. The ownership is shared by farmers, INERA, Monsanto, and the cotton companies. Since 1999 and after the privatization in 2005, cotton producers have been shareholders of the three cotton companies. The cotton sector has since been jointly owned by the Burkina Faso Government, the private sector, and the farmers. So, all the stakeholders can benefit from any technology such as the Bt cotton recently adopted and commercialized.

But for sustainability, many efforts should be made by all the stakeholders in order to meet the future challenges and strengthen the cotton sector. To avoid challenges related to benefit sharing, there is a need to establish trust in both the public and private sector, as a means to secure the future of agbiotech public/private partnerships (PPPs) in the country through transparent interactions and clearly defined project priorities, roles and responsibilities among core partners (Obidimma et al. 2009; Obidimma and Abdallah 2012). There is also a need for improved communication strategies and appropriate media response, to obviate unwarranted public perceptions of the project. The country should continue building its capacity in terms of human resources and laboratory equipments, to respond adequately to the new challenges.

Studies should continue to take place on issues like the resistance of targeted pests to Bt toxins, and the environmental risks related to Bt cotton.

Scientists should continue training the extension staffs of the cotton companies as well as farmers on technical issues about cotton as a whole but particularly about Bt cotton. Farmers' associations still need support to be empowered in some issues such as communication, management, and negotiation. Farmers should also be sensitized, trained, and convinced about the implementation of refuge zones in their Bt cotton fields (20 % of the area planted). Continued monitoring will be required to determine the technical and economic viability of Bt cotton over the short and long term. Since cotton is part of a production system including cereals, studies should be undertaken on the development of a suitable farm council which will enhance the ability of the producer to make the diagnosis of his/her farm, in order to consider actions to improve incomes.

During this phase of deployment of the product, it is still critical to have an integrated communication and awareness training program for all players on the product life-cycle, including researchers, developers, cotton company extension workers, seed producers, farmers, and staff at the ginning facilities.

Stewardship awareness and training were essential thanks to the FARA-SABIMA project, not only for staff and stakeholders involved with early-stage research CFTs, but also all along the development chain through to farmers commercializing Bt cotton (Traoré and Héma 2011).

At this stage of commercialization of Bt cotton, it is important to prevent the mixing of GM and conventional seed. For example, during this process of conducting trials in farmers' fields when it became clear that INERA scientists were unable to personally oversee all plantings, monitoring, harvesting, and ginning activities for the cotton grown by the 20 farmers, it was decided that the solution was to adapt its approach to more fully engage its partners and other stakeholders along the entire value chain, and to serve as a training and facilitating organization.

We identified critical control points (CCPs), and developed standard operating procedures (SOPs) along the cotton production process from taking seed to farmers through to ginning, to ensure product integrity and prevent inadvertent mixing of seeds (Traoré and Héma 2011). Tracking of product and verification procedures are essential to reduce the risk of cross-contamination.

Efforts towards the intensification of production systems must continue to ensure better profitability, and INERA scientists have a great role to play by training farmers on the technical itinerary of production of Bt cotton, in collaboration with the cotton companies and Monsanto.

A challenge is about to be overcome, but other challenges including weeding, hunger, and malnutrition should be considered, since biotechnology can offer some solutions. Indeed, in Burkina Faso, hunger still remains in some regions, and every day, if one considers the number of Burkinabè who are going to sleep without eating, this serves as extra motivation to continue the effort. At INERA, the responsibility of the scientific community is to help farmers by providing them access to nutrition. Biotechnology can be used as a tool to increase productivity and to target malnutrition by improving food quality (biofortification). Research in biotechnology can target other issues such as biotic and abiotic stresses (resistance to insects and diseases, drought tolerance), and tissue culture. So, some biotechnology projects such as the African biofortified sorghum project, the Bt cowpea project, and the SABIMA-FARA project are in the pipeline with hopes of success. INERA is still in discussion with Monsanto for the evaluation of the Roundup Reddy Flex herbicide-tolerant cotton. Hopefully, the fight will continue to reduce poverty and help farmers achieve food security.

Acknowledgments The Government of Burkina Faso, which allowed all the processes including special authorization to import and experiment with transgenic cotton; CNRST and INERA; the Burkina Faso cotton industry, and Monsanto, for financial support for the experiments. Thanks to Dr. Roger Zangré, Pr. Alassane Séré, and all other members of the cotton research team of INERA (Dr. Déhou Dakuo, Dr. Doulaye Traoré, Dr. Oula Traoré, Mr. Denys Sanfo, Dr. Bazoumana Koulibaly, Mr. Gaspard Vognan, Mr. Hughes Somé), and all the technical and administrative staff for providing information. Africa Harvest and the Rockefeller Foundation for organizing the Bellagio writing retreat.

References

Abdallah N (2010) GM crops in Africa: challenges in Egypt. GM Crops 1(3):116–119. http://www.landesbioscience.com/journals/gmcrops/03Abdallah2GMC1-3.pdf

African Biosafety Network of Expertise (ABNE) (2010) Global modern biotechnology adoption trend. http://www.nepadbiosafety.net/subjects/socioeconomic-issues/adoption-trend

Chetaille A (2006) Evaluation des besoins en renforcement de capacités sur la biosécurité en Afrique de l'Ouest. In: Etude pour le ministère des Affaires étrangères. Groupe de recherche et d'échanges technologiques (GRET), Paris, 65pp

Compaoré B (2004) Speech to the opening plenary by his excellency Blaise Compaore of Burkina Faso at the ministerial conference on harnessing science and technology to increase agricultural productivity in Africa: West African perspectives. Ouagadougou, 21–23 June 2004. http://www.fas.usda.gov/icd/stconf/event6/event6compaore

Club du Sahel (2005) Importance économique et sociale du cotton en Afrique de l'Ouest: Rôle du cotton dans le développement, le commerce et les moyens d'existence. Secrétariat du Club du Sahel et de l'Afrique de l'Ouest/OCDE. Paris, 72pp

Greenplate JT, Mullins JW, Penn SR, Dahm A, Reich BJ, Osborn JA, Rahn PR, Ruschke L, Shappley ZW (2003) Partial characterization of cotton plants expressing two toxin proteins from *Bacillus thuringiensis*: relative toxin contribution, toxin interaction, and resistance management. J Appl Ent 127:340–347

Héma SAO (2004) Contribution à la caractérisation biochimique de la résistance de Helicoverpa armigera (Hübner, 1808) (Lepidoptera, Noctuidae) aux pyréthrinoïdes au Burkina Faso. Mémoire de D.E.A., Ecole Doctorale Régionale de Biotechnologie, Université de Ouagadougou, 37pp

Héma SAO, Konaté G, Traoré O, Menozzi P (2009a) Biochemical characterization of the cotton bollworm *Helicoverpa armigera* resistance to pyrethroids in Burkina Faso. Pak J Biol Sci 12(13):964–969

Héma SAO, Somé HN, Traoré O, Greenplate J, Abdennadher M (2009b) Efficacy of transgenic cotton plant containing the Cry1Ac and Cry2Ab genes of *Bacillus thuringiensis* against *Helicoverpa armigera* and *Syllepte derogate* in cotton cultivation in Burkina Faso. Crop Prot 28:205–214. doi:10.1016/j.cropro.2008.09.014

INSD (2008) Recensement général de la population et de l'habitation de 2006: résultats définitifs. Institut National de la Statistique et de la Démographie, Ministère de l'Economie et des Finances. Ouagadougou, 52pp

Institute of Environment and Agricultural Research (INERA) (2002) Description de la filière cotonnière au Burkina Faso. Rapport intermédiaire de l'Institut de l'Environnement et de Recherches Agricoles (INERA), campagne 2002–2003, Ouagadougou, 34pp

Institute of Environment and Agricultural Research (INERA) (2007) Production cotonnière au Burkina Faso: la recherche, un maillon essentiel. Programme Coton/INERA, Dossier, Eurêka 50:11–53

International Food Policy Research Institute (IFPRI) (2006) Regional strategic alternatives for agricultural led growth and poverty reduction in West Africa. A final draft report, 96pp. http://ebrary.ifpri.org/cdm/ref/collection/p15738coll2/id/127175

International Service for the Acquisition of Agri-biotech Applications (ISAAA) (2007) Bt maize trials perform positively in Egypt (Crop Biotech Update). Retrieved 8 Oct 2012, from http://www.isaaa.org/kc/cropbiotechupdate/article/default.asp?ID=795

International Service for the Acquisition of Agri-biotech Applications (ISAAA) (2010) Field visit to Bt maize fields in Egypt (Crop Biotech Update). Retrieved 10 Aug 2012, from http://www.isaaa.org/kc/cropbiotechupdate/article/default.asp?ID=6587

James C (2010) A global overview of biotech (GM) crops: adoption, impact and future prospects. GM Crops 1(1):8–12

Martin T, Ochou GO, Hal NF, Vassal JM, Vaissayre M (2000) Pyrethroid resistance in the cotton bollworm, *Helicoverpa armigera* (*Hübner*), in West Africa. Pest Manag Sci 56:549–554

Michel B, Togola M, Téréta I, Traoré NN (2000) Cotton pest management in Mali: issues and recent progress. Cahiers Agric 9:109–115

Ministère de l'Agriculture, des Ressources Hydrauliques et de la Pêche (MAHRH) (2008) Document guide de la révolution verte. Ministère de l'Agriculture, de l'Hydraulique et des Ressources Halieutiques. Ouagadougou, 97pp

National Centre for Scientific and Technological Research (CNRST) (1995) Plan stratégique de la recherche scientifique: Synthese générale. Centre National de la Recherche Scientifique et Technologique, Ministère des Enseignements Secondaire, Supérieur et de la Recherche Scientifique, Ouagadougou, 75pp

Obidimma CE, Abdallah SD (2012) Building trust in biotechnology crops in light of the Arab Spring: a case study of Bt maize in Egypt. Agric Food Secur 1. http://www.agricultureandfoodsecurity.com/content/1/S1/S4/abstract. doi:10.1186/2048-7010-1-S1-S4. Accessed 08 Oct 2012

Obidimma CE, Fiona T, Jim L, Abdallah D, Peter S (2009) A social audit model for agro-biotechnology initiatives in developing countries: accounting for ethical, social, cultural, and commercialization issues. J Technol Manag Innov 4(3):24–33

Penn JB (Personal communication) (2004, June 21–23) Prepared remarks for closing press briefing at the ministerial conference on harnessing science and technology to increase agricultural productivity in Africa: West African perspectives. Ouagadougou, http://www.fas.usda.gov/icd/stconf/event6/event6compaore.html, 26 Sept 2012

Perlak FJ, Oppenhuizen M, Gustafson K, Voth R, Sivasupramaniam S, Heering D, Carey B, Ihrig RA, Roberts JK (2001) Development and commercial use of Bollgard cotton in the USA—early promises versus today's reality. Plant J 27(6):489–501

Pray CE, Huang J, Hu R, Rozelle S (2002) Five years of Bt cotton in China—the benefits continue. Plant J 31(4):423–430

Sarant L (2012) Bt maize seed shipment seized in Egypt: Egypt's legal battle to regulate Monsanto's GMOs. Egypt Independent. Retrieved 10 Aug 2012, from http://www.gmwatch.org/latest-listing/1-news-items/14087-bt-cotton-seed-shipment-seized-in-egypt

Sawahel W (2008) Egypt approves commercialization of first GM crop. Retrieved 10 Aug 12, from http://www.scidev.net/en/news/egypt-approves-commercialisation-of-first-gm-crop.html

Séré A (2007) Les bouleversements en cours dans la production agricole. L'Afrique va-t-elle encore courir derrière une nouvelle révolution verte? Biotech Echo 4:1–2

SOFITEX (2012) État des superficies en coton conventionnel et transgénique. Direction du développement de la production cotonnière, Campagne 2012/2013

Traoré FB (2007) Nous n'avons rien à perdre en adoptant les OGM. Interview par Hamed Traoré du Président de l'Union Nationale des Producteurs de Coton du Burkina, Dossier, Eurêka 50:50–51

Traoré H, Héma SAO (2011) Case study Burkina Faso: conducting Bt cotton controlled release field trials with farmers in Burkina Faso. In: Johnson L et al (eds) Agricultural biotechnology in Africa: stewardship case studies. Forum for Agricultural Research in Africa (FARA), Accra, pp 56–59

Traoré O, Denys S, Vitale J, Traoré K, Bazoumana K (2008) Economics & marketing: testing the efficacy and economic potential of Bollgard II under Burkina Faso cropping conditions. J Cotton Sci 12:87–98

Vitale J, Glick H, Greenplate J (2006). The economic impacts of Monsanto's Bollgard II in West Africa: Emprirical evidence from Burkina Faso. A paper presented at the 10th ICABR Conference, Ravello

Vitale J, Glick H, Greenplate J, Abdennadher M, Traoré O (2008) Second-generation Bt cotton field trials in Burkina Faso: analyzing the potential benefits to West African farmers. Crop Sci 48:1958–1966

Vitale JD, Vognan G, Ouattara M, Traoré O (2010) The commercial application of GMO crops in Africa: Burkina Faso's decade of experience with Bt cotton. AgBioForum 13(4):320–332

Zangré GR (2009) Du protocole de Cartagena à l'exploitation à grande échelle du coton Bt au Burkina Faso. Biotech Echo 27:1–3

Chapter 3
Opportunities and Challenges of Commercializing Biotech Products in Egypt: Bt Maize: A Case Study

Shireen K. Assem

Abstract GM technology has been developed to solve the problems of hunger and poverty, and also to create job opportunities and improve the quality of life in developing countries. In agriculture, genetically modified organisms (GMOs) are developed to possess several desirable traits, such as resistance to pests, herbicides, and harsh environmental conditions, improved product shelf life, increased nutritional value, and possession of traits for production of valuable goods such as drugs (pharming). Egypt was one of the few countries to realize in the 1980s the importance of GM crops in achieving sustainable agriculture. Technology transfer and building capacities for the development of agricultural crops through biotechnology started in the early 1990s. In 2008, Egypt approved the cultivation and commercialization of a Bt maize variety, marking the first legal introduction of GM crop into the country. The case of commercialization of Bt maize in Egypt is very unique. It has been 5 years since the first approval for commercialization of this biotech product in Egypt; it is important to discuss the current situation of this biotech product in the Egyptian market and evaluate the benefits for the farmers and the consumers after the adoption of this product. It is also important to highlight the constraints in commercializing this product, and the opportunity for adoption and commercialization of other biotech products in the Egyptian market in the future.

Keywords Bt maize • GM maize • GMO • Agricultural biotechnology • Public–private partnership • Egypt

Abbreviations

EPA Environmental Protection Agency (USA)

S.K. Assem (✉)
Agricultural Genetic Engineering Research Institute, Cairo Giza 12619, Egypt
e-mail: shireen_assem@yahoo.com

F. Wambugu and D. Kamanga (eds.), *Biotechnology in Africa*,
Science Policy Reports 7, DOI 10.1007/978-3-319-04001-1_3,

An Overview of Agriculture in Egypt

Egypt lies in the northeastern corner of Africa, with a total land area of 1 million square kilometers. It is bounded by the Mediterranean Sea to the North, the Red Sea to the East, and Sudan to the South. The river Nile, the longest river in the world, provides the critical water supply to this arid country. Egypt's population is about 83 million, with over 97 % of the population concentrated in about 4 % of the total area, while the remaining 96 % is uninhabited. This high population is concentrated in the Nile Valley and Delta.

From the earliest days of history, Egypt has always been a predominantly agricultural country, which created most of Egypt's wealth. For over 5,000 years, the farmers created a civilization based on the union of the land and the Nile River. It was one of the earliest civilizations, and it had a profound influence on the region. Grain, vegetables, and fruit were grown, cattle, goats, pigs, and fowl were raised, and fish from the Nile were caught, and eventual surpluses, after deduction of the various taxes, were sold in the markets. Due to the yearly inundations, the soil remained fertile but agricultural techniques were not very efficient. Improvements were rare, implements remained primitive, and the breeding of better livestock was haphazard. Fishing appears to have existed on a very small scale. But practically all of the fish consumed were caught from the Nile. A large part of the manufactured goods came from the families which produced the raw materials.

Today, agriculture in Egypt combines the use of traditional methods with a rich base of knowledge of the land and the environment. Agriculture is a major economic issue in Egypt as it provides a source for local food supply, for international trade, for balance of payments, land and water use, and as a basic product for food and fiber manufacturing. Hence, every aspect of the economic structure of the country relates to agriculture. Banking, transportation, tax and tariff structures, subsidies, and local and international markets are all part of the agricultural system of the country; not to mention politics, of course.

Only 3 % of the land, equivalent to approximately 2.5 million hectares, is devoted to agricultural production. Within this area, old lands represent 66 %, the new lands represent 32 % and the rain-fed area represents only 2.4 %. This leads to the country having one of the world's lowest levels of cultivable land per capita.

Agriculture's contribution to GDP is gradually diminishing, but it is still an important activity. Agriculture is still considered a principal sector in the economy, contributing about 13 % of the GDP and providing close to 30 % of employment. About 90 % of the agricultural land is in the Nile Delta, and the remaining is within a narrow strip along the Nile, between Aswan and Cairo. The rich, cultivated land, irrigated by the Nile is very fertile and allows double cropping. Nevertheless, the small area that can be cultivated, as well as problems related to salinity and water, results in Egypt being dependent on imports for about half of its food supply (Karembu et al. 2009).

The principal crops are wheat, rice, maize, and sugarcane. The government policy is to enhance agriculture as a major contributor to the national economy, by promoting privatization and decreasing government control and subsidies. The major challenges for agricultural development in Egypt are the limited arable land base, erosion of land resources, loss of soil fertility, salinity, and the high rate of population growth of 1.9 % (Elbanna 2011).

Despite the small area of arable land and insufficient water supply, Egypt's agricultural sector remains one of the most productive in the world. Moreover, reclaiming Egypt's desert lands has been a major government objective for more than 50 years. Plans are underway to reclaim 3.5 million acres by the year 2017, including the South Valley Development Project's attempts near Lake Nasser, in addition to the 1 million acres of desert that have been already reclaimed in the past years.

The Egyptian revolution on the 25th of January, 2011, was a major turning point for the whole society. There are a number of challenges for the development of different sectors in Egypt. The greatest challenge lies within the agriculture sector, one of the most important sectors impacting the economy in the country.

Maize in Egypt

Maize (*Zea mays* L.) is one of the most important cereal crops in Egypt after wheat and rice. Egypt produces about 6 mm tons of corn every year, on approximately 750,000 ha of land. Egypt is self-sufficient in its white corn needs, while it produces only a small fraction of the quantity required to meet its need for yellow corn. Egypt is the fourteenth largest producer of corn in the world, with an average corn consumption of 12 million tons, making it the eighth largest consumer of corn in the world, and the fifth largest importer of corn in the world, with an import average of 5 million tons of yellow maize annually, valued at US $1.6 billion. As of 2010, according to FAO statistics, maize was recorded as the second largest imported commodity in Egypt after wheat.

Maize probably yields more industrial products than any other grain crop. The stalks, leaves, cobs, and grain all have some commercial value, although that of the kernels is by far the greatest. All of the maize that is planted in Egypt is intended to ultimately provide food for the human population, but the route from the field to the human stomach is somewhat varied. Some of the maize grain produced in Egypt is consumed directly by humans. A large share is fed to animals and a small share is fed to poultry (Krenz et al. 1999).

In recent years, there has been an increase in demand for maize grain from the bread industry, where 20 % of white maize grain is mixed with 80 % of wheat flour to reduce the imported quantity of wheat. With the increase in rice and cotton cultivated areas in 2011–2012, a significant decrease occurred in the acreage of corn planted.

Biotechnology Programs and Maize Research

Maize production in villages and farms faces many constraints such as pests, weeds, environmental degradation, soil nutrient depletion, and low fertilizer inputs. Science and biotechnology are the most promising recourse to alleviating most of these constraints.

The agronomy research on maize at the Agricultural Research Center is focusing on the improvement of both yield and resistance to major pests, primarily late wilt and downy mildew diseases, as well as corn borers. There has been the development of new high-yielding yellow and white hybrids (three-way and single-cross hybrids), and recently, more efforts have been channeled into breeding for drought and heat tolerance, to derive new inbreds tolerant under such stress conditions. Moreover, studies are undertaken on the socio-economic impact affecting farmers' adoption of new technology (especially new hybrids) through on-farm research trials and field questionnaires.

Research for the development of genetically modified (GM) crops is ongoing in a number of research institutions and universities in Egypt. The GM crop research is being carried out on different crops and specific traits, depending on the economic importance and nature of constraints in production and utilization of these crops.

Genetic engineering programs started in Egypt in 1990 by the initiation of the National Agricultural Genetic Engineering Laboratory (NAGEL), which was renamed the Agricultural Genetic Engineering Research Institute (AGERI) in 1992 with presidential declaration.

AGERI presented a model of moving research into commercial application through the successful interaction between scientists at AGERI and the University of Wyoming, who have been involved in collaborative research studies for 6 years on Bt. The research efforts led to the development of a biological pesticide based on a highly potent strain of Bt isolated from the Nile Delta. This strain is extremely effective against a broad range of insects: Lepidoptera (moths), Coleoptera (beetles), and Diptera (mosquitoes). An additional significant feature of this strain is its capacity to kill nematodes.

AGERI has successfully managed to manufacture its first biopesticide *Agerin* based on the insecticidal bacterium *Bacillus thuringiensis*. Agerin is capable of protecting a broad range of important agricultural commodities, of controlling a number of biomedically significant pests, and has the potential for sales on a worldwide scale. AGERI succeeded in establishing a commercial business entity for the commercialization of research results conducted in AGERI and to sell AGERI products (Madkour 2000).

Moreover, biotechnology and genetic engineering research activities were conducted at AGERI on different crops: potato, tomato, cotton, maize, cucurbits, wheat, banana, etc. These crops were developed with desirable characteristics such as biotic stress resistance and abiotic tolerance. However, the developed plants did not reach the stage of commercial release due to the lack of national legislation on biotech crops, despite the fact that Egypt has ratified the Cartagena Protocol. The development of solid policies, legislation, and guidance relating to the safe use of the GM products at the national level is important to maintain human health as well as a safe environment. However, public awareness is the biggest challenge when it comes to the scientific development of GM crops for commercialization.

Corn Borers Affecting Maize

In Egypt, infestation with the corn borers *Sesamia cretica Led., Lep., Noctuidae, Ostrinia nubilalis* (*Hübner*), *Lep., Pyralidae* and *Chilo agamemnon Bles., Lep., Pyralidae*, cause serious economic yield loss by boring both the stem and the ear (Semeada et al. 1999). These borers are also considered the principal cause of fungal and bacterial diseases. *Sesamia cretica* is considered to be the most serious of the borers. This species attacks maize plants shortly after emergence, devours the whorl leaves, and may kill the growing point of apical shoot tip, causing what is known as a "dead heart". It is also capable of damaging older plants and excavating tunnels into the stem, ears, and/or cobs. This pest lays eggs during March, so it causes complete death of small maize plants in April and May, leading to a decrease in population and drastic yield losses. Control of *S. cretica* in maize fields is commonly done by the application of chemical insecticides, either as sprays or granules, directly to the whorl. Side-effects of this chemical control on the agroecosystem include the destruction of natural enemies of pests, outbreaks of mite population, and environmental pollution (Soliman 1994). Moreover, recommendations have been made by the Ministry of Agriculture for the farmers to cultivate maize during the period between late May and mid-June, as this greatly helps to avoid the risk of damage by corn borers, especially *Ostrinia nubilalis*.

Bt Maize

In Egypt, there are about 20 million poor people; most of the country's rural poor live in Upper Egypt. Since farming is the most important source of income and sustenance in rural areas, about three-quarters of the population of rural areas are small-scale farmers. Small farmers (less than half a hectare) do not control stem borers, because the damage caused by the caterpillars is hidden and difficult to detect, and the costs of conventional pesticides, environmentally friendly biological compounds, and organic chemicals treatment are high. Small-scale farmers who do spray often risk exposure to the chemicals because they use unsuitable equipment and/or fail to use protective clothing.

GM maize provides a new management tool for small-scale farmers, has the potential to increase yield where the stem borer is a problem, and decrease the need for chemical applications (Pilcher et al. 1997).

In 2008, the Egyptian Ministry of Agriculture approved decisions made by the National Biosafety Committee (NBC) and the Seed Registration Committee to allow for commercialization of a GM Bt corn hybrid (Ajeeb YG). This Bt corn (Mon810) is a GM yellow grain hybrid developed by Monsanto. This hybrid produces a protein throughout the corn plant that is toxic to certain *lepidopteran* insect species. The resulting plant is one of several transgenic corn varieties generally referred to as Bt corn, or Bt maize. The substance produced through the genetic alteration is identified as δ-endotoxin or the Cry1Ab protein.

The commercial release of Bt maize in Egypt was not the first approval for a biotech crop commercialization in Africa. Earlier in 1997, the Department of Agriculture in the Republic of South Africa (RSA) issued the first conditional commercial release permits for GM crops. These were for GM cotton and maize. To date, RSA has commercially approved herbicide-tolerant maize, soybean and cotton, as well as insect-resistant maize and cotton. Therefore, South Africa was the first country in Africa to commercially grow a staple food of its population—white maize—in GM form. Moreover, in 2008, after Egypt, Burkina Faso approved the commercial release of Bt cotton, making it the third country to approve the commercial release of a GM crop in Africa. Currently in Burkina Faso, out of a total of 424,810 ha containing cotton plantations in the country, 247,000 ha or 58 % comprises the Bt cotton crop. Therefore, by the approval of Bt maize for commercialization in Egypt, Egypt became the second among the African countries, and the first country from the Arab world, to approve the cultivation of GM crops. Egypt has always had a leading role within the Arab world and among the Middle Eastern countries.

Approval of Transgenic Bt Maize for Commercialization

Bt maize that has been approved for commercialization in Egypt is a high-yielding yellow single cross. Its leaves stay green after seed maturity, and are used for the production of silage. It has been approved for cultivation and commercialization as animal feed only and not for human consumption.

It took 10 years to receive the necessary approvals after completing the required risk assessment tests. In 1999, Monsanto initiated a joint project with the private Egyptian company Fine Seeds International for the development, commercialization, and distribution of Bt maize in Egypt. From 2005 to 2008, the NBC led the risk-assessment and testing process of the Bt maize (Karembu et al. 2009; Adenle 2011). In 2008, Bt maize was approved for commercial use, making Egypt the first country in the Arab world to commercialize a biotech crop (Chalony and Moisseron 2010). The endorsement was based on a series of field trials conducted between 2002 and 2007 for the variety MON810 hybrid, produced by a biotechnology company. Bt crops produce a toxin that guards against pests. The variety to be distributed is a cross between MON810 line, and a yellow maize line, with resistance to three corn borer pests, developed by scientists of Monsanto in South Africa. The Cairo-based company "Fine Seeds International S.A.E." is partnering with Monsanto to distribute the Bt maize hybrid in Egypt.

Every year, the local seed company distributing the Bt maize undergoes monitoring for the product and submits an annual report to the NBC, including the quantity imported, the quantity sold, the places where it is sold, and the names of the traders. The report also includes the efficiency of the GM corn and the bioassay with corn borers, the refuge, and any adventitious presence.

The Cultivation of Bt Maize in Egypt

In 2008, Egypt planted 700 ha (1,729 acres) of Bt maize (James 2010). The area cultivated with Bt corn in 2010–2011 was about 1,700 ha (4,201 acres), of which about 1,000 ha were cultivated in the new project "Toshka" in Southern Egypt, and gave almost a double yield, compared to the yield of the conventional hybrid crop. There were 50 t of Bt seeds imported between May 2011 and December 2012 from South Africa, and about 56 t imported between 2012 and 2013, from the same source. Farmers would like to grow biotech corn, since they know that it gives them higher yields and uses less fertilizer, pesticides, and water. However, the availability of Bt corn seed is still very limited (Mansour 2012).

Benefits of Biotech Maize in Egypt

About 13.3 million farmers in 25 countries have planted biotech crops spread across 125 million hectares. Of these farmers, over 90 % or 12.3 million are small and resource-poor farmers from developing countries. The high adoption rate reflects the fact that biotech crops have consistently performed well and delivered significant economic, environmental, health, and social benefits to both small and large farmers (Navarro 2009). Egypt is one of three African countries, together with South Africa and Burkina Faso, that knew early on the importance of producing GM crops. Since 1960, food production in Egypt has failed to keep pace with the increasing rate of consumption in the country. This gap could be attributed to the low level of investment in agricultural research, the slow growth in agricultural products, and the rapid increase in per capita food consumption (Abdallah 2010).

Experience has shown significant benefits of growing Bt corn. Bt corn can make the farmer's life easier, allow more targeted use of pesticides, and improve the quality and quantity of the harvest.

A local distributor, Dr. Adel Yasin, the owner of Fine Seeds Co., said that, after intensive Bt maize field trial studies in over 36 maize-growing areas in Egypt, the use of biotech maize was proved to save on pesticide usage, give almost 100 % protection from stem borers, and increase yield by 30–40 % above the conventional maize varieties. In the first 3 years after approval, Egypt imported about 30–40 t of Bt maize, which has been grown on an area of about 1,500–2,000 acres each year. In 2011, Egypt increased the imported amount to 50 t. During these 4 years, farmers expressed their satisfaction with biotech maize, noting that they benefited more than when they use conventional varieties. One of the maize farmers, Alsayed, planted 1 acre of Bt maize to compare with the conventional maize that he normally grew: "I found a 25 % increase in yield and high maize quality in the Bt maize variety compared to the conventional variety, and although I planted it late, the Bt maize was able to resist borer infestation" he says. Morsy, another maize farmer, said that he sprayed the conventional maize variety three times with pesticides, which cost him about US $90, while he used no pesticides for the Bt maize variety (Karembu et al. 2009). The only comment on that variety as noted by Dr. Adel Yasin was that, 10 years ago when the company started the process of issuing the necessary approvals for the commercialization of this GM corn, the conventional variety "Ajeeb" was one of the best superior varieties produced in Egypt in terms of field performance and yield at that time. However, now, after 10 years, the Field Crops Research Institute at the Agricultural Research Center (ARC), and other seed companies in Egypt, including Fine Seeds itself, have produced other yellow maize varieties by conventional breeding with higher yield and better field performance (with respect to traits other than insect-resistance) varieties. These varieties now are competing with Bt maize as higher-yielding maize.

According to Elbanna (2011), developers of Bt maize reported the following economic benefits in 2009: (1) an increase in yield per hectare resulting in a gain of US $267, plus (2) an insecticide saving equivalent to US $89 per hectare for a total gain of US $356 per hectare. (3) minus the additional cost of seed per hectare at US $75, for (4) a benefit per hectare of US $281. Extrapolating from these data, the benefit from planting 2,000 ha in 2010 is of the order of US $550,000.

Additionally, the use of Bt maize in Egypt would have an import substitution value, from increased self-sufficiency of maize, plus savings of foreign exchange.

Biosafety System in Egypt

A national biosafety system is an important element of the national strategy to develop biotechnology products. It is also instrumental in involving different stakeholders in the definition of the national project, and in facilitating collaboration with foreign countries (Chalony and Moisseron 2010).

Egypt currently has no official biosafety legislation, though a regulatory framework exists. The existing framework follows the Cartagena Protocol on Biosafety, and encompasses ministerial decrees regulating the registration of GM varieties.

The Egyptian National Biosafety Committee (NBC) was created in January 1995. Its members are representatives from the ministries of agriculture, health, industry, higher education and scientific research, and foreign affairs, the Egyptian Environmental Affairs Agency (EEAA), and non-governmental organizations (NGOs). They include policy makers, experts from universities and research centers, and non-technical members, all headed by the Minister of Agriculture. A series of ministerial decrees from March 1995 to August 1997 established the rules of Egyptian biosafety (Chalony and Moisseron 2010).

The NBC is in charge of establishing policies and procedures regarding biotechnology across the whole country. It is the official body responsible for ensuring that biotechnology continues to be safe, and facilitating access to modern biotechnology that is generated abroad. All organizations that undertake biotechnology research must establish an Institutional Biosafety Committee (IBC) to organize the safety of their activities. These committees report back to the NBC. If an import permit is required, the NBC notifies a secondary specialized agency (for example, the Supreme Committee on Food Safety), before authorizing field tests. Once the authorization is given, a team of NBC inspectors carries out the monitoring.

Foreign donors have been strongly involved in setting up the biosafety system in Egypt. The Agricultural Genetic Engineering Research Institute (AGERI) has also played a central role in building institutional capacities for biosafety (Chalony and Moisseron 2010).

In the beginning of May 2011, and after the Egyptian revolution, the newly appointed Minister of Agriculture issued a new ministerial decree reconstituting the Egyptian National Biosafety Committee (NBC).

Commercialization and the Current Status of Bt Maize

Although the commercialization of Bt maize in Egypt appears to have been successful, it did not meet a number of goals. First, Bt maize was only commercialized for animal feed and silage, and not for human consumption. Second, it was only allowed to introduce the gene to yellow maize, which, in comparison to white maize, is unpopular and not widely grown. Third, the adoption rate of Bt maize in Egypt has been slow, and it is still held back by trust challenges (Obidimma and Abdallah 2012). While white maize is consumed locally in bread–making, yellow maize is utilized for animal feed, and Egypt imports 5 million tons of yellow maize annually, at much cheaper prices than of the crop produced locally.

Since its approval for commercialization in 2008, Bt maize has raised a lot of controversy and opposition. The opposition has sometimes been from the media, the public, or the traders, and sometimes even from other scientists who work in other fields of science, and also from members of the government itself.

On the 8th of March 2012, the Minister of Agriculture ordered a temporary suspension of planting of MON810, in reaction to false information circulated regarding health concerns, which have no apparent scientific basis. In the current political environment in Egypt, government officials appear particularly sensitive to criticism, even when untrue (Mansour 2012).

Biosafety of Bt Maize

As with any other GM crop, the safety of Bt corn has been thoroughly assessed, both during development, and on an ongoing basis in the areas in which it is grown. The process of developing and marketing a genetically enhanced crop is carefully regulated, and national regulatory bodies have concluded that Bt corn is safe for food, feed, and for the environment.

Bt maize is extensively cultivated, mainly in the USA—on almost 24 million hectares, in 2011. It has been approved for cultivation there since 1995. Bt maize is also used to an appreciable extent in Argentina, South Africa, Canada, and the Philippines. Worldwide, the total area under GM maize is 51 million hectares; about three-quarters of this is maize with one or more types of resistance against pests—i.e., Bt maize.

In Europe, cultivation of Bt maize is concentrated in Spain. Growing on an area of 97,000 ha (2011), it represents 26.5 % of the Spanish maize production. Farmers have also shown Bt maize in Portugal, the Czech Republic, Romania, Poland, and Slovakia, although on considerably smaller acreage.

Genetically modified (GM) maize MON810 (notification reference C/F/95/12-02) was authorized under Directive 90/220/EEC (EC 1990) in the European Union (EU), for all uses (with the exception of food uses) by the Commission Decision 98/294/EC (EC 1998). A final consent was granted to the applicant (Monsanto Europe S.A.), by France, on 3rd August, 1998. Food uses of maize derivatives were notified according to Article 5 of the Novel Food Regulation (EC) No 258/97, on 6th February 1998 (The EFSA Journal 2012).

MON810 was subject to regulation primarily by the Environmental Protection Agency (EPA) under the Federal Insecticide, Fungicide, and Rodenticide Act (FIFRA), 7 U.S.C. §§ 136-136y; under the Federal Food, Drug and Cosmetic Act (FFDCA), 21 U.S.C. §§371-379d; by the Animal and Plant Health Inspection Service (APHIS) under the Federal Plant Pest Act (FPPA), 7 U.S.C. §§ 150aa-150jj; and by the Plant Quarantine Act (PQA), 7 U.S.C. §§ 151-164a, 166–167 (as amended).

EPA issued an Experimental Use Permit for field testing MON810; and it later registered MON810 for commercial sale and use, subject to a time limit, specified conditions (which subsequently have been strengthened), and exempted the pesticidal portion from the requirement of having a residue limit (tolerance) in food. APHIS authorized the field testing of MON810, and subsequently granted it non-regulated status, i.e., APHIS determined that MON810 is not subject to APHIS' regulatory oversight based on current knowledge. APHIS conducted an

Environmental Assessment under the National Environmental Policy Act (NEPA), 42 U.S.C. §§ 4321-4370e, on the basis of which it issued a finding of no significant impact on the environment (Finding No Significant Impact, FONSI), and also concluded that there were no issues under the Endangered Species Act (ESA), 16 U.S.C. §§ 1531–1544 (Case study II).

In 2009, the European Commission, and the Panel on Genetically Modified Organisms of the European Food Safety Authority (EFSA GMO Panel), reported that the information available for maize MON810 addresses the scientific comments raised by member states, and that maize MON810 is as safe as its conventional counterpart with respect to potential effects on human and animal health. The EFSA GMO Panel also concludes that maize MON810 is unlikely to have any adverse effect on the environment in the context of its intended uses, especially if appropriate management measures are put in place in order to mitigate possible exposure of non-target *Lepidoptera* (The EFSA Journal 2009).

Following the request from the EFSA GMO Panel, assessing the monitoring report for the 2010 growing season of maize MON810 provided by Monsanto Europe S.A., on 7th September 2011, the EFSA GMO Panel adopted a scientific opinion on the 2009 monitoring report of maize MON 810.

The EFSA has found that Monsanto's MON810 GM maize poses no risk to human health or the environment, based on data from the 2010 growing season. The scientific opinion on the Post-Market Environmental Monitoring (PMEM) report for 2010 concluded that the cultivation of the maize MON810—otherwise known as YieldGuard—had no adverse effects on human and animal health or the environment.

EFSA said that the latest opinion corroborates its previous assessment on maize MON810 for cultivation during the 2009 season. The report stated: "*From the data submitted by the applicant in its 2010 monitoring report, the EFSA GMO Panel does not identify adverse effects on the environment, human and animal health due to maize MON810 cultivation during the 2010 growing season*" (The EFSA Journal 2009). On the other hand, in September 2012, Séralini et al. published online in the scientific journal *Food and Chemical Toxicology*, an article describing a 2-year feeding study on rats, investigating the health effects of GM maize NK603, with and without Roundup WeatherMAX®, and Roundup® GT Plus alone (both are glyphosate-containing plant protection products). EFSA was requested by the European Commission to review this publication, and to identify whether clarifications are needed from the authors.

The EFSA, concluded that the Séralini et al. study, as reported in the 2012 publication, does not impact the ongoing re-evaluation of glyphosate, and does not see a need to reopen the existing safety evaluation of maize NK603 and its related stacks. EFSA will give the authors of the Séralini et al. (2012) publication the opportunity to provide further information on their study to EFSA.

It is important to mention that, following the request by the applicant for the renewal of the authorization for placing maize MON810 in the market, the EFSA GMO Panel adopted a scientific opinion on the renewal under Regulation (EC) No. 1829/2003 of maize MON810 for import, processing for food and feed uses, and cultivation, in June 2009 (EFSA 2009). The EFSA GMO Panel concluded that "maize MON810 is unlikely to have any adverse effect on the environment in the context of its intended uses, especially if appropriate management measures are put in place in order to mitigate possible exposure of non-target (NT) *Lepidoptera*" (EFSA 2012).

Factors Limiting the Adoption of Biotech Crops

In general, the State and its public research policies play an important role in involving and mobilizing the private sector, including foreign partners. The biotech corn had strong opposition when it was introduced in Egypt. The adoption and commercialization of biotech products in Egypt is facing a number of challenges.

The greatest are the absence of the "Biosafety Law" and the fluctuation in political will towards GMO adoption and commercialization. These two factors reflect negatively on the mutual trust between the government and the public including farmers, consumers, media, traders, and investors. Other factors including the economy, environment, and education, in addition to dissemination of social and political information, are playing a critical role in the negative attitude towards biotech crops as well.

The management of biotechnologies involves many participants of a well-planned strategy, which associates and merges the interests of several actors. Various sectors of the society have a vital role, and they must cooperate in influencing the adoption of biotech products.

The Role of the Government

In the past decade, Egypt has declared and supported science and technology, but the amount of public investment in this area was minimal compared to other countries; the amount of its GDP spent on research was 0.2 % in 2010.

After Egypt's revolution, science and education are slowly emerging from the post-revolution chaos as national priorities. Revitalizing Egypt's sclerotic and chronically underfunded research, education and innovation systems will require sweeping reforms and substantial rises in spending. On 1st June, 2011, the Egyptian cabinet approved the first post-revolution budget, which boosted science despite the severe social and economic crises gripping the country. Research spending increased from 2.4 L.E. billion (US $404 million) to 3 billion L.E. in the 2011–2012 financial year. The education budget also jumped, by 16 %, to 55.7 billion L.E. The increase in science spending still leaves it at only around 0.4 % of the GDP, much less than the 1–2 % that researchers say they would like. The goal is to reach that level within 4 years (Nature 2011).

The Role of the Legislative Authority

The legislative authorities in Egypt should recognize that life sciences, especially biotechnology, is growing very fast all over the world, and Egypt must be ready to catch up with this growth and be a partner and not just a consumer. Therefore, to move forward and to seize the opportunity in harnessing the potential of biotech crops, a comprehensive Biosafety Law compatible with the Cartagena Protocol should be put in place to foster the safe use and management of GMOs.

The Role of Scientists

Biotechnology research is mainly directed to improve the livelihood of the poor and solve problems associated with our daily life. As the number of biotechnology products is increasing and more biotech crops are being released and adopted in several countries, public concern with regard to biotic crops is also increasing. More effort should be put in with regard to public awareness. Communication with media and decision makers is a very important activity for scientists. A scientist should work closely with public and deliver scientific data in a very simple way (Navarro 2009). Communication could be through seminars, conference, and workshops for the public, farmers, media people, and decision makers. Moreover, open field trials could be performed with contribution of representatives from different sectors including farming and trading, to raise awareness about Biotech products.

Education channels on Egyptian television could be a good way to deliver simplified information about biotechnology to the public. In 2011, Egypt launched the Agriculture television channel that works closely with ARC and the extension sector. This TV channel could also be the link to farmers and public to raise awareness toward Biotech products.

The Role of the Media

The communication practitioner's role in the biotechnology area is a significant one. Surveys show that much of the information that consumers have of science, and to make sense of scientific breakthrough, is based on what they gain from newspapers, television, the radio, and the internet (Navarro 2009). Few people have direct experience with agricultural biotechnology or even direct exposure to scientists and researchers working in the field. Hence, media play a crucial role in providing people with the information necessary to make decisions about technology options and their potential risks and benefits. Another important role for media is that they allow citizens to gauge the climate of opinion around them, which in turn influences what people will think about a certain issue (Scheufele 2007). The public should be well-informed and educated about the adoption and production of a GMO, before its introduction. Advice should be sought from indigenous scientists and should not be based on foreign scientists only, before adoption of GMO. GM crop trials should be conducted before approval, and the public should be informed.

The Egyptian Biotechnology Information Center (EBIC), which was established in ARC in collaboration with the International Service for the Acquisition of Agri-Biotech Applications (ISAAA) in 2003–2004, could play a very important role in delivering clear and simple information in brief to the public, in the form of regular monthly reports, in addition to annual reports. It responds to specific information needs, promotes and advances a broader public understanding of crop biotechnology, and monitors the local agri-biotech environment.

Conclusion

Commercializing a GM crop for the first time in a developing country like Egypt is a big challenge, and some issues related to transparency and clearness should be considered. The most important is transparency between the partners. All information related to the process of production and commercialization should be available for all partners at the same level. The public should be well-informed and educated about benefits of the adoption and production of GM crops in their country, before their introduction. Advice should be sought from indigenous scientists, and should not be based only on what foreign scientists have to say, before adoption of the GM crop. GM crops field trials should be carried out before approval, and the public should be informed. Media people and farmers should participate in the field trials for the evaluation of the GM crop. More attention should be drawn towards the benefits of GM production for large-scale and small-scale farming, and benefits for the consumers.

Acknowledgments The author is grateful to each of the reviewers who contributed substantial time and effort to this study. Special thanks to Prof. Hanaiya El-Itriby for comments on the draft of the manuscript.

References

Abdallah N (2010) GM crops in Africa: challenges in Egypt. GM Crops and Food 1(3):116–119
Adenle AA (2011) Adoption of commercial biotech crops in Africa. IPCBEE 7:176–180
Case study No. II. Bt maize. www.whitehouse.gov
Chalony C, Moisseron J (2010) Research governance in Egypt. Sci Technol and Soc 15 (2):371–397
Egypt invests in its science: Latest budget establishes research as a national priority (2011) Nature 474 266. Retrieved from www.nature.com
Elbanna H (2011) Terza Giornata Mondiale Del Mais, Bioenergy, Italy. (CREMONA, 18–20 MARZO 2011), Maize Production in Egypt. Hany Elbanna, Chief Editor of Al-Ahram Agriculture Magazine

European Food Safety Authority (2009) Scientific opinion on applications for renewal of authorization for the continued marketing of maize MON810 and existing derived food and feed products. EFSA J 1149:57–85

European Food Safety Authority (2012) Scientific opinion on the annual monitoring report by Monsanto on the cultivation of GM maize MON 810 in 2010. EFSA J 10(4):2610

James C (2010) Global status of commercialized biotech/GM crops. ISAAA Brief No.42. International Service for the Acquisition of Agri-Biotech Applications, Ithaca

Karembu M, Nguthi F, Ismail H (2009) Biotech crops in Africa: the final frontier. ISAAA AfriCenter, Nairobi, pp 1–34

Krenz RD, El Guindy M, Ariza-Nino E, Siddik I (1999) Utilization of maize in Egypt. Report No.72. Retrieved from http://pdf.usaid.gov/pdf_docs/Pnacs287.pdf

Madkour MA (2000) Egypt: biotechnology from laboratory to the marketplace: challenges and opportunities. In: Persley GJ, Lantin MM (eds) Agriculture biotechnology and the poor: proceedings of an international conference, Washington, DC: 21–29 Oct 2000

Mansour S (2012) Egypt grain and feed annual report, wheat and corn production on the rise (USDA Foreign Agricultural Service). Retrieved from http://gain.fas.usda.gov

Navarro MJ (2009) Communicating crop biotechnology: stories from stakeholders. ISAAA Brief No.40. International Service for the Acquisition of Agri-Biotech Applications, Ithaca

Obidimma CE, Abdallah SD (2012) Building trust in biotechnology crops in light of the Arab Spring: a case study of Bt maize in Egypt. Agriculture & Food Security, 1(Suppl 1):S4. Retrieved from http://www.agricultureandfoodsecurity.com/content/pdf/2048-7010-1-S1-S4.pdf

Pilcher CD, Rice M, Obrycki J, Lewis L (1997) Field and laboratory evaluation of transgenic *Bacillus thuringiensis* corn on secondary Lepidopteran pests (*Lepidoptera*: *Noctuidae*). J Econ Ent 90:669–678

Review of the Séralini et al (2012) publication on a 2-year rodent feeding study with glyphosate formulations and GM maize NK603 as published online on 19 September 2012 in food and chemical toxicology. EFSA Journal 10(10):2910

Scheufele D (2007) Opinion climates, spirals of silence and biotechnology: public opinion as a heuristic for scientific decision-making. In: Brossard D, Shanahan J, Nesbitt TC (eds) The media, the public and agricultural biotechnology. CAB International, Cambridge, MA, pp 231–244

Semeada AM, Fédière G, Elkheikh MAK, El-Sherif SI (1999) Current status of corn borers in maize fields and pilot utilization of *Sesamia crefica* granulovirtis "sscgv"as a biocontrol agent. J Agric Sci Mansoura Univ 24(11):6955–6964

Séralini GE, Claira E, Mesnagea R, Gress S, Defargea N, Malatestab M, Hennequinc D, de Vendômoisa JS (2012) Long term toxicity of a roundup herbicide and a roundup-tolerant genetically modified maize. Food Chem Toxicol 50(11):4221–4231. doi: http://dx.doi.org/10.1016/j.fct.2012.08.005

Soliman M (1994) Corn borers affecting maize in Egypt. In "insect resistant maize, recent advances and utilization". In: Proceedings of an international symposium, CIMMYT, 27 Nov–3 Dec 1994

Chapter 4
Genetically Modified Crops Commercialized in South Africa

E. Jane Morris and Jennifer A. Thomson

Abstract Genetically modified (GM) crops have been in commercial production in South Africa since 1997, when Bt cotton and maize were approved by an advisory committee acting under interim legislation. The Genetically Modified Organisms (GMO) Act was passed in 1997, but only implemented in 1999. The Act was modified in 2006 to bring it into line with the requirements of the Cartagena Protocol on Biosafety. The area planted to GM crops has steadily increased, with the majority of maize, soybeans and cotton being GM. Problems with field-resistance to Bt in *Busseola fusca* in maize started in 2007, linked to very low levels of compliance among farmers in planting refugia. This continues to be a problem. Smallholder GM maize farmers in KwaZulu–Natal have been planting this crop since 2001, and have experienced higher yields and other benefits. South Africa is experiencing a number of delays with approvals of new GM crops, particularly those developed, at least in part, in this country. Additionally, the Biotechnology Innovation Centers established after the publication of the National Biotechnology Strategy in 2001 have been closed and incorporated into a new Technology Innovation Agency (TIA). As TIA has been largely dysfunctional since 2010 this has created a hiatus in funding. As a result of these problems South Africa is currently at a crossroads in terms of the development and adoption of GM crops. For the country to move forward, bold steps are required, and a number of recommendations are listed.

The African Centre for Gene Technologies is a collaborative initiative of the Universities of Pretoria, Witwatersrand and Johannesburg, with the Agricultural Research Council and the CSIR.

E.J. Morris (✉)
African Centre for Gene Technologies and Department of Biochemistry, Faculty of Natural and Agricultural Sciences, University of Pretoria, P. O. Box 75011, Lynnwood Ridge, 0040 Pretoria, South Africa
e-mail: ejanemorris@gmail.com

J.A. Thomson
Department of Molecular and Cell Biology, University of Cape Town, Private Bag Rondebosch, Cape Town, South Africa
e-mail: Jennifer.thomson@uct.ac.za

F. Wambugu and D. Kamanga (eds.), *Biotechnology in Africa*, Science Policy Reports 7, DOI 10.1007/978-3-319-04001-1_4,

Keywords South Africa • Genetically modified (GM) crops • Bacillus thuringiensis • Herbicide tolerance • Maize • GMO Act

Abbreviations

ARC	Agricultural Research Council (South Africa)
BIC	Biotechnology Innovation Center
Bt	*Bacillus thuringiensis*
CSIR	Council for Scientific and Industrial Research (South Africa)
DEA	Department of Environment Affairs (South Africa)
DST	Department of Science and Technology (South Africa)
DTI	Department of Trade and Industry (South Africa)
EC	Executive Council (South Africa)
GM	Genetically modified
GMO Act	Genetically Modified Organisms Act (South Africa)
GMOs	Genetically modified organisms
HT	Herbicide tolerance
ISAAA	International Service for the Acquisition of Agri-Biotech Applications
PTM	Potato tuber moth
RSA	Republic of South Africa
SAGENE	South African Committee on Genetic Experimentation
TIA	Technology Innovation Agency (South Africa)

Introduction

Genetically modified (GM) crops first obtained commercial regulatory approval in South Africa in 1997 when Monsanto's Bt cotton and MON810 maize (Yieldgard), both with insect-resistance, were approved for general release under the interim legislation involving the advisory committee SAGENE. SAGENE was mandated to carry out biosafety risk assessment reviews, and the final approval resided with the Directorate of Plant and Quality Control, of the Department of Agriculture. Permits were linked to the import of seed, and were issued under the authority of the Agricultural Pests Act of 1983. Following the issuing of permits, the crops were commercially planted in the 1998–1999 growing season.

The Genetically Modified Organisms Act (GMO Act) (Act no.15 of 1997) was passed in 1997 but was only implemented in 1999 following the promulgation of regulations. Since that date, the structures put in place under the GMO Act have had overall authority in terms of release of GMOs. The Act was modified by the subsequent GMO Amendment Act (Act no. 23 of 2006) to bring it into line with the requirements of the Cartagena Protocol on Biosafety. The National Environmental Management Biodiversity Act (Act no. 10 of 2004) also affects decisions made on release of GMOs by including a provision to require an environmental assessment if there is a threat to any indigenous species or the environment.

The GMO Act makes provision for an Executive Council to make decisions on all applications regarding GMOs. This Executive Council is comprised of members from all relevant and interested government departments, which are required to reach a consensus before any new applications are approved.

In 2005, the Executive Council decided to suspend all current and new applications requesting commodity clearance approval for the import of GM maize. This was based on concerns raised by the Department of Trade and Industry (DTI) with regard to the extent to which such approvals might disadvantage local producers. A study subsequently confirmed the benefits to the country if domestic production of approved GM maize events were allowed, and commodity clearance approvals eventually resumed in 2011, but with strengthened regulatory control measures for GM commodities.

Growth of Commercialized Crops

The area planted to GM crops in South Africa has steadily increased since they were first introduced. White maize, used for human consumption, covered approximately 1.64 million hectares in 2012, and of this 80.5 % was GM with 38 % of the plants containing the Bt gene, 12 % herbicide tolerance (HT), and 50 % stacked with both traits (James 2012).

Yellow maize, used for animal fodder and chicken feed, covered approximately 1.19 million hectares. Of this, 93 % was GM, with the percentages of the three available traits being 31 %, 21 %, and 48 % respectively (James 2012).

Herbicide-tolerant (HT) soybeans accounted for 90 % of the total crop of 450,000 hectares in 2012 (James 2012).

Table 4.1 shows the GMO events approved for general release in South Africa from 1997 until 2012. There were no general release approvals in 2013.

It appears that adoption of currently approved traits is reaching saturation. This is because not all plantings require Bt insecticide-resistance since, in many cases cost savings can be achieved by applying fungicide and insecticide simultaneously through overhead irrigation, when needed. In addition, some regions are not subject to severe stalk borer pressure (James 2011). However, new traits in the pipeline, such as fungal-resistance and drought tolerance, may serve to further enhance adoption levels.

Cotton production has declined in recent years due to movement away from risky dryland regions, to regions under irrigation, where it has to compete with maize or soybeans. There have also been problems with cotton gin closures. Therefore, only about 11,000 hectares were planted in 2012, but of that, 95 % contained the stacked Bt and herbicide-resistant genes while the remaining 5 % was herbicide-resistant used as refugia (James 2012).

Table 4.1 GMO general release approvals under the GMO Act, 1997

Event	Crop	Trait	Company	Year approved
TC1507	Maize	Insect-resistant Herbicide-tolerant	Pioneer	2012
Bt 11xGA21	Maize	Insect-resistant Herbicide-tolerant	Syngenta	2010
GA21	Maize	Herbicide-tolerant	Syngenta	2010
MON89034xNK603	Maize	Insect-resistant Herbicide-tolerant	Monsanto	2010
MON89034	Maize	Insect-resistant	Monsanto	2010
Bollgard IIxRR flex (MON15985x MON88913)	Cotton	Insect-resistant Herbicide-tolerant	Monsanto	2007
MON88913 (RR flex)	Cotton	Herbicide-tolerant	Monsanto	2007
MON810xNK603	Maize	Insect-resistant Herbicide-tolerant	Monsanto	2007
Bolgard RR	Cotton	Insect-resistant Herbicide-tolerant	Monsanto	2005
Bollgard II, line 15985	Cotton	Insect-resistant	Monsanto	2003
Bt11	Maize	Insect-resistant	Syngenta	2003
NK603	Maize	Herbicide-tolerant	Monsanto	2002
GTS40-3-2	Soybean	Herbicide-tolerant	Monsanto	2001
RR lines 1445 & 1698	Cotton	Herbicide-tolerant	Monsanto	2000
Line 531/Bollgard	Cotton	Insect-resistant	Monsanto	1997
MON810/Yieldgard	Maize	Insect-resistant	Monsanto	1997

Problems with the Development of Insect Resistance

Bt maize has been grown in South Africa on over 200 million hectares since 1996, although the first commercial release was in 1998. This represents one of the largest selection experiments for the development of insect-resistance to this toxin. The first reports of field-resistance in *Busseola fusca*, one of the most important lepidopteran pathogens of maize, came in 2007 from irrigated maize in the Northern Cape Province (Van Rensburg 2007). The fact that resistance was initially relatively low is rather surprising, as the time involved exceeds the time for resistance to develop to most conventional insecticides (Bates et al. 2005).

Field-evolved resistance is defined as a genetically-based decrease in susceptibility of a population to a toxin caused by exposure of the population to the toxin in the field (Tabashnik et al. 2009).

The way field-resistance is managed in Bt crops is by ensuring that the dose of the Cry protein is high, and by planting refuges of non-Bt plants to ensure pest survival. The concept behind the high dose strategy is that the rate of resistance evolution is driven primarily by the frequency and survival of heterozygotes (RS, where R = resistant phenotype; and S = sensitive phenotype), so the goal is a dose that will kill them according to the following (where LD_{99} is the dose that will kill 99 % of the population) (Environmental Protection Agency 1998):

High dose = 25 × LD_{99} of SS
(i.e., 25 times stronger than the dose that is lethal for 99 % of SS pests)

The concept behind refugia is that with the high dose, very few RR moths will survive in a Bt crop, but many SS moths will emerge from the non-Bt refugia plants. Thus all RR moths will mate with SS moths, resulting in RS moths which will be killed by the high dose plants. In order for this to occur there must be sufficient refugia plants, and they must be planted in such a way that encourages SS moths to emerge. The refugia requirements specify either a 20 % refuge planted to conventional maize, which may be sprayed with non-Bt insecticides, or a 5 % refuge area that may not be sprayed.

The 2007 report of resistance was followed up by a survey conducted among 80 farmers at the irrigation scheme (Kruger et al. 2009). Results demonstrated that initially there was a very low level of compliance to the establishment of refugia. Those farmers that planted refugia tended to establish them outside the irrigated area. Ovipositing moths are known to select plants growing under higher moisture conditions. Although farmers had signed contracts committing to the planting of refugia, initial compliance appeared to be very low (7.7 % in 2008).

A recent study (Kruger et al. 2012) indicated that although compliance with refugia requirements has improved significantly, insect-resistance is considerably more widespread than initially thought within the main maize-growing area of the country. There are still problems with compliance and uncontrolled use of insecticidal sprays, showing irresponsible management of GM crop technology by farmers, chemical, and seed companies.

The trouble in developing countries is that there is reluctance among farmers to set aside land that will be planted to insect-susceptible plants when they are paying for insect-resistant seeds, and it is difficult to enforce compliance. Perhaps one of the solutions is to mix 20 % of refugia seeds in a bag. This makes it easy for the company and there is no need for compliance monitoring. Studies have shown, however, that for some pests, interplant movement by larvae would render this strategy less effective. Larvae may actively avoid feeding on Bt plants, or larvae developing on non-Bt plants may move to toxic plants and die, thus reducing the effective size of the refuge (Bates et al. 2005). Currently all refugia planting is structured such that the refuges are separate from Bt plants. The best hope for the future is probably the introduction of stacked insecticidal genes targeting different receptor sites.

GM Maize: The Experiences of Smallholder Farmers

A study has been conducted into the experience of smallholder GM maize farmers in KwaZulu–Natal over a period of eight seasons (2001/2002–2009/2010). This is the only example to date and internationally, where a subsistence crop that is also a staple food has been produced by small-scale farmers using GM seed. The study answers the question: "taking into consideration the immense variability in production conditions between seasons and in production practices between farmers, have smallholders benefited from the adoption of insect-resistant (Bt) and/or herbicide-tolerant (HT) maize seed?" (Gouse 2012).

The results show that smallholder Bt adopters enjoyed higher yields than their conventional maize-planting counterparts, and in most seasons, were better off despite paying more for their seed and not saving on insecticides, since insecticide usage in the smallholder environment was in any case limited. In the case of HT maize, farmers also benefited through higher yields brought about by more effective chemical weed control, compared to the manual weed control practices of conventional maize planting farmers. The total labor saving benefits of HT maize, however, depended on farmers' production systems. When farmers made use of "planting-without-ploughing" i.e., no-till or minimum-till production, they used significantly less labor than when they ploughed land in preparation for planting with tractors or oxen (Gouse 2012).

In most seasons studied, HT maize produced the highest net farm income, and farmers planting these seeds were more efficient than both conventional maize farmers and Bt farmers. "Farmers seem to be willing to pay for the weed control convenience of HT maize and based on adoption figures, farmers value the yield-increasing and labor-saving benefits of HT maize higher than the borer control insurance of Bt maize" (Gouse 2012). These findings are in direct contradiction to claims that "the types of GM crops and traits currently on the market are widely acknowledged... to offer little to small-scale farmers in the developing world" (Glover 2008).

Some Problems with Crops in the Pipeline

One of the problems associated with approvals of GM crops under the GMO Act of 1997 is that the Executive Council responsible for these decisions comprises members from six different departments, all having different agendas. These six departments are the Department of Agriculture Fisheries and Food, Department of Science and Technology (DST), Department of Trade and Industry (DTI), Department of Health, Department of Environment Affairs (DEA), and Department of Labor. Dissent by any of these departments will result in the rejection of a permit application. When one looks at the list of commodity clearances (excluding general release clearances) from 2001 to 2012, there is a complete hiatus from 2004 to 2011 due to the concerns of the DTI alone. Similarly, the DEA has been instrumental in either delaying or preventing approval of several applications.

Below are two problems which are examples of the type of obstacles scientists face when trying to obtain permission for various applications.

African Biofortified Sorghum

More than half a billion people around the world rely on sorghum as a dietary staple. Its tolerance for drought and heat make it an important food crop in Africa (it is indigenous to Ethiopia and Sudan). However, it lacks certain essential nutrients. In order to give it added nutritional value, the African Biofortified Sorghum project is being run by an international consortium under the leadership of Africa Harvest, an African-based international nonprofit organization. African biofortified sorghum contains the gene for a high-lysine storage protein from barley and has increased levels of Vitamin A, iron, and zinc.

In 2006, scientists from the Council for Scientific and Industrial Research (CSIR) in South Africa, applied to the Registrar of the Directorate for Genetic Resources Management in the National Department of Agriculture, the body that administers the GMO Act of 1997 (as amended in 2010), to undertake greenhouse trials. This application was denied by the Executive Council on a number of grounds.

(i) In view of the potential risks pertaining to environmental impact (as a result of gene flow), the Council recommended that this experiment be conducted on a non-indigenous species with no wild relatives in South Africa.
(ii) Taking into consideration the Council's concerns about gene flow, the applicant should take note that the possibility of obtaining a trial release or general release authorization with this species – as with any other indigenous species – would be extremely low.
(iii) The Council expressed concerns regarding the current containment levels of the facilities that would be involved in the proposed activities, and indicated that such activities should be conducted in at least a Level 3 containment facility.

In its appeal, the CSIR pointed out that the South African Biotechnology Strategy stresses the importance of value-addition to indigenous crops. In sharp contrast, the decision by the EC could be interpreted to mean that no research on indigenous crops should be allowed.

The CSIR also noted that the EC was prejudging future applications for trials and/or general release. Why turn down an application for a greenhouse trial on the supposition that at some time in the future, an application might be made for a trial or general release, and that permission for this might not occur? Decisions by the EC should be made on scientific grounds.

Finally, the appeal noted that the CSIR did, indeed, have a Level 3 containment facility that had been approved by the Directorate for Genetic Resources. A clear case of the left hand not knowing what the right hand was doing.

Two appeals were turned down, but finally, in 2009, permission was granted. However, the damage had already been done. Most of the R&D for this project was moved to Kenya, where approval for GM sorghum greenhouse trials was obtained within 3 months, and trials began within 5 months.

This outcome shows that South Africa, which has the greatest expertise and capacity in plant biotechnology in Africa, is likely to lose the advantage for carrying out projects that involve applications for permits under the GMO Act. This is due to the uncertainty of the regulatory goals and the lengthy process that each application requires. These types of projects will in the future most likely be funded and initiated in other African countries (Thomson et al. 2010).

Insect-Resistant Potatoes

The larvae of the potato tuber moth (PTM), *Phthorimaea operculella*, bore into potato leaves, stems, and tubers, causing extensive damage. In addition, fungi and mites can grow in the galleries formed by the PTM's burrowing, resulting in the decomposition of the tuber. The impact of the PTM fluctuates from season to season in response to climate, but recurs regularly at high levels, and can cause up to R 40 million in losses per annum (Visser and Schoeman 2004).

In July 2008, an application was submitted by the Agricultural Research Council (ARC) of South Africa to the Directorate of Biosafety of the Department of Agriculture, Forestry, and Fisheries, for a general release of GM potato event SPUNTA-G2. This event had been developed by Michigan State University and carried the Bt *Cry1Ia1* gene (Douches et al. 2002). The required information was submitted, including socio-economic impact data, and a stewardship plan. However, on August 25, 2009, the application was rejected. The reasons for this refusal included:

- No evidence that other pest management strategies against PTM have been considered or compared with the release of GM Spunta.
- The capacity of small-scale farmers to implement risk-management measures could potentially be onerous.
- Considering the biology of potatoes, vegetative material (tubers) may be used for propagation, which may complicate risk management.
- PTM is not a major pest for stored potatoes compared with rodents.

These issues were addressed in a reply from the ARC dated September 21, 2009, requesting to appeal the decision.

1. Information on many of the socio-economic issues can only be collected if the application is approved. This approval is needed, to enable the farmer

participatory evaluation, which must precede any decision on whether the ARC will use this trait for the improvement of South African potato varieties. Indeed, farmer participatory trials will help to answer many of the questions regarding the impact of the trait on potato production and on farmers, as posed by the Executive Council in its decision.
2. A general release approval for Spunta-G2 is essential for farmers to undertake assessments regarding productivity, production constraints, appearance, taste, storages, and marketability.
3. The use of vegetative planting material requires no additional effort compared to the use of true seed with other crops.
4. The levels of all potato pests vary from season to season, but PTM remains the primary storage pest. "The Prokonnuus 2006–2007 data collected at fresh produce markets in South Africa clearly indicate that tuber moth was the third major cause of spoilage in potatoes, after greening and mechanical damage, and caused million of Rands in losses".

From the above, it could be argued that the Executive Council overstepped its mandate when they determined that smallholder farmers would not need this technology. It is the mandate of the ARC and farmers themselves to assess whether this GM technology is appropriate for local use. Weak decision-making processes have jeopardized the funding for this and other public sector projects (Thomson et al. 2010).

According to the regulations governing the GMO Act, appeals should be heard within 180 days. This appeal was lodged in 2009 and the appeal board was appointed within the required time, but the appeal board did not reach a decision for more than a year. The minutes of the Executive Council meeting of July 2012 indicate that a decision has been reached, but the nature of the decision has not been communicated to the applicant, and was not in the public domain at the time of going to press. The project was officially cancelled at the end of September 2012, in the absence of a regulatory decision after 3 years of waiting (M. Koch, personal communication 2012).

Support for GM Crop Development

Donor organizations have funded, and continue to fund, various GM projects in South Africa. However, support from South African sources is also important for long-term continuity and success. Following the publication of the National Biotechnology Strategy of 2001, the DST established a number of Biotechnology Innovation Centers (BICs), including PlantBio, the BIC responsible for disbursing funding for projects involving plant biotechnology. While there has generally been a favorable attitude towards the development of GMOs, the funding allocation for GM projects has been limited. In 2010, the BICs were incorporated into the newly created Technology Innovation Agency (TIA), which has a mandate to promote commercialization of all technologies, not only biotechnology. The fact that TIA has been largely dysfunctional has created a hiatus in funding, and has particularly had a negative impact on the biotechnology sector, which has been seen as too long-term and risky compared with other technologies.

Another initiative funded by the DST in terms of the National Biotechnology Strategy is Biosafety SA. The establishment of a National Biosafety Platform that could provide regulatory guidance and support for GM product development was identified as a priority in 2004. As a first step, Mr. Willy de Greef, a world-renowned authority on biosafety and regulatory issues, was commissioned to investigate the matter. He submitted a memorandum entitled "Creation of a Biosafety and Regulatory Consortium in South Africa" in which he proposed the creation of a center of excellence for biosafety research.

Biosafety SA was established in 2008 to provide the following offerings:

- Guidance and assistance to academia and companies on GMO regulatory requirements
- Supporting strategic GMO biosafety research
- Facilitation and management of regulatory compliance projects
- Information management and dissemination
- Capacity building and training
- Decision support services for regulators

Biosafety SA originally fell under PlantBio. With the merging of the BICs into TIA, Biosafety SA currently forms part of TIA, though this situation is currently under review.

The role of the platform in the commercialization pipeline of biotechnological products was originally seen as crucial, and requiring strategic investment. The platform should co-invest in such projects for future commercial return, which lead to long-term financial sustainability. Unfortunately, it seems that there are currently no GM products in the pipeline requiring support for regulatory compliance. Contributing factors to this situation are: (1) most academic work is in the early stage, (2) later-stage projects are mainly undertaken through public-private partnerships with multinationals, where the multinationals have their own regulatory expertise, and (3) local industry is unwilling to accept the market risk associated with a new GM product.

Although Biosafety SA appears to play a valuable support role in some of the other areas within its mandate, the fact that their support for commercialization is not required represents a telling indictment of the state of GMO development within South Africa.

Public Attitudes

GM food is widely consumed in South Africa, although levels of awareness amongst the general population are low, and there is widespread indifference (Racovita et al. 2013). Organizations such as the biotechnology stakeholders' association AfricaBio, and the government funded program on Public Understanding of Biotechnology, have made some inroads into providing information to the public. However, much still remains to be done.

Meanwhile, a number of anti-GM groups are actively lobbying against GMOs. Some are stirring up opinion at the parliamentary level and advocating a total ban on GM crops. They also present frequent challenges to the decision-making processes under the GMO Act, aimed at destabilizing the biosafety process.

It can be largely attributed to anti-GM activists that South Africa is now faced with onerous GM labeling requirements under the Consumer Protection Act. Despite existing legislation promulgated by the Department of Health that only required labeling of GM food in specific cases, such as when allergens are present, the regulations promulgated by the DTI under the Consumer Protection Act would require labeling of all products that contain more than 5 % GM ingredients or components. This is not likely to be practical to enforce, and would require elaborate and costly testing to ensure adherence.

The Way Forward

South Africa is currently at a crossroads in terms of the development and adoption of GM crops. Despite widespread adoption to date, problems continue to emerge. Insect-resistance prejudices the efficacy of first-generation GM crops. Concerns around socio-economic issues are hampering decision-making. Government departments have differing attitudes towards GMOs, making it difficult to reach a consensus. Inefficiencies and lack of expertise in the government lead to long delays and poor decisions.

For the country to move forward, some bold steps are required. Some specific recommendations include:

- The recently published Bio-economy Strategy (DST 2013) mentions the role of genetic engineering as a critical technology for agriculture. The role of GMOs in contributing to agriculture, food security, and economic development must be recognized and endorsed at the highest levels of government, with buy-in from all relevant government departments.
- The decision-making mechanisms under the GMO Act should be revised as follows:
  - Executive Council members and Advisory Committee members should be obliged to attend regular training courses to ensure consistency and quality in biosafety reviews and decision-making. This should not be optional.
  - An initial framing step should be included in the review process, in line with international trends. This would ensure that all parties are in agreement concerning the issues at stake. This framing step should include an agreement on benefits as well as risks to be considered (Morris 2011).

- The Executive Council should not attempt to duplicate the work of the Advisory Committee, but should respect the assessment of risk provided by the Advisory Committee.
- There should be more clarity as to the specific types of data that an applicant must provide (e.g., feeding trials to be carried out, etc.). The requirements should reflect the level of risk, and not be so onerous that public sector projects are unable to comply.
- If socio-economic considerations are to be taken into account, this should only be at the stage of general release (not contained use or field trials), and should be evidence-based, not conjectural.
- A decision should require only a clear majority vote. A single government department should not have the right of veto.
- Time frames specified in the GMO Act and Regulations should be strictly adhered to, and the applicant should have right of recourse if a response is not forthcoming within the specified time frames. Appeals, in particular, should be concluded in 3 months.
- Systems should be put in place to ensure that conditions specified in the release permit are adhered to (e.g., maintenance of refugia).

Acknowledgments The authors gratefully acknowledge the very useful inputs provided by Ms. Muffy Koch.

References

Bates SL, Zhao JZ, Roush RT, Shelton AM (2005) Insect resistance management in GM crops: past, present and future. Nat Biotechnol 23:57–62

Douches DS, Li W, Zarka K, Coombs J, Pett W, Grafius E, El-Nasr T (2002) Development of Bt-cry5 insect resistant potato lines Spunta-G2 and Spunta-G3. Hortic Sci 37:1103–1107

DST (2013) The bio-economy strategy. ISBN 978-1-919966-06-09. 44pp. http://www.dst.gov.za/images/ska/Bioeconomy%20Strategy.pdf

Environmental Protection Agency (1998) Final report of the FIFRA scientific advisory panel subpanel on Bacillus thuringiensis (Bt), plant and resistance management, meeting held on Feb 9–10, 1998. Environmental Protection Agency, Washington, DC. http://www.epa.gov/scipoly/sap/meetings/1998/february/finalfeb.pdf

Glover D (2008) Made by Monsanto: the corporate shaping of GM crops as a technology for the poor. STEPS working paper 11. STEPS Centre, Brighton

Gouse M (2012) Farm-level and socio-economic impacts of a genetically modified subsistence crop: the case of smallholder farmers in KwaZulu–Natal, South Africa. Doctoral dissertation, University of Pretoria

James C (2011) Global status of commercialized biotech/GM crops: 2011. ISAAA Brief No. 43. ISAAA, Ithaca

James C (2012) Global status of commercialized biotech/GM crops: 2012. ISAAA brief no. 44. ISAAA, Ithaca

Kruger M, Van Rensburg JBJ, Van den Berg J (2009) Perspective on the development of stem borer resistance to Bt maize and refuge compliance at the Vaalharts irrigation scheme in South Africa. Crop Prot 28(8):684–689

Kruger M, Van Rensburg JBJ, Van den Berg J (2012) Transgenic Bt maize: farmers' perceptions, refuge compliance and reports of stem borer resistance in South Africa. J Appl Entomol 136: 38–50

Morris EJ (2011) A semi-quantitative approach to risk–benefit assessment. Transgenic Res 20: 1055–1071

Racovita M, Obonyo DN, Abdallah R, Anguzu R, Bamwenda G, Kiggundu A, Maganga H, Muchiri N, Nzeduru C, Otadoh J, Rumjaun A, Suleiman I, Sunil M, Tepfer M, Timpo S, van der Walt W, Kaboré-Zoungrana C, Nfor L, Craig W (2013) Experiences in sub-Saharan Africa with GM crop risk communication. GM Crops Food Biotechnol Agric Food Chain 4(1):1–9

Tabashnik BE, van Rensburg JBJ, Carrière Y (2009) Field-evolved insect resistance to Bt crops: definition, theory, and data. J Econ Entomol 102(6):2011–2025

Thomson JA, Shepherd DN, Mignouna HD (2010) Developments in agricultural biotechnology in sub-Saharan Africa. AgBioForum 13(4):314–319

van Rensburg JBJ (2007) First report of field resistance by the stem borer, *Busseola fusca* (Fuller) to Bt transgenic maize. S Afr J Plant Soil 24:147–151

Visser D, Schoeman AS (2004) Flight activity patterns of the potato tuber moth, *Phthorimaea operculella* (*Lepidoptera*: *Gelechiidae*). Afr Entomol 12:133–137

Part II
Building the Bio-economy and Commercialization Challenges

Chapter 5
Moving Africa Towards a Knowledge-Based Bio-economy

E. Jane Morris

Abstract The development of a bio-economy in Africa will require enhanced bio-innovation for sustainable development. It will also require structural and policy changes to ensure that the systems are in place to support the bio-economy. Already, many developed and developing countries are placing emphasis on the development of a bio-economy, and African countries must not be left behind.

There are many bio-innovation success stories in Africa, some of which are highlighted in this paper, such as the development of cereals resistant to parasitic infection, banana tissue culture, veterinary vaccines and diagnostics, microbial fermentation products, and value addition to primary agricultural products through agro-processing.

There are also challenges to overcome, including access to finance and business skills, legislative and intellectual property hurdles, and the need for enhanced technical capacity. Some of these challenges are being overcome in novel ways, such as through regional collaborations and public–private partnerships, but strong leadership by Africa's politicians and a change of mindset regarding the importance of science, technology and innovation are needed if the opportunities presented by the bio-economy are to be fully realized.

Keywords Bio-economy • Innovation • Technology transfer • Agriculture • Genetically modified (GM) crops

The African Centre for Gene Technologies is a collaborative initiative of the Universities of Pretoria, Witwatersrand, and Johannesburg, with the Agricultural Research Council and the CSIR.

E.J. Morris (✉)
African Centre for Gene Technologies and Department of Biochemistry, University of Pretoria, PO Box 75011, Lynnwood Ridge, Pretoria 0040, South Africa
e-mail: ejanemorris@gmail.com

F. Wambugu and D. Kamanga (eds.), *Biotechnology in Africa*, Science Policy Reports 7, DOI 10.1007/978-3-319-04001-1_5,

Abbreviations

AATF	African Agricultural Technology Foundation
ABSP	Agricultural Biotechnology Support Project
ABSPII	Second phase of ABSP
AHBFI	Africa Harvest Biotech Foundation International
ARC	Agricultural Research Corporation (Sudan)
ASARECA	Association for Strengthening Agricultural Research in East and Central Africa
AU/PANVAC	Pan-African Veterinary Vaccine Centre of the African Union
BecA	Biosciences East and Central Africa
BIC	Biotechnology Innovation Centre (South Africa)
Bio-EARN	Eastern Africa Regional Program and Research Network for Biotechnology, Biosafety and Biotechnology Policy Development
Bio-Innovate	Bio-resources Innovations Network for Eastern Africa Development
CGIAR	Consultative Group on International Agriculture Research
CIMMYT	CGIAR International Maize and Wheat Improvement Center
CORAF/WECARD	West and Central African Council for Agricultural Research and Development
CSIR	Council for Scientific and Industrial Research (South Africa)
EPHTFCP	Eastern Province Horticultural Traditional Food Crop Project (Kenya)
FAO/UNIDO	Food and Agriculture Organization/United Nations Industrial Development Organization
FARA	Forum for Agricultural Research in Africa
FTO	Freedom-to-operate
GM	Genetically modified
HCDA	Horticultural Crop Development Authority (Kenya)
ICRISAT	International Crops Research Institute for the Semi-Arid Tropics
IFAD	International Fund for Agricultural Development
IITA	International Institute for Tropical Agriculture
IP	Intellectual property
IPRs	Intellectual property rights
KARI	Kenya Agricultural Research Institute
KHCP	Kenya Horticultural Competitiveness Project
KIRDI	Kenya Industrial Research Development Institute
MDG	Millennium Development Goal
NEPAD	New Partnership for Africa's Development
OECD	Organization for Economic Co-operation and Development
PIPRA	Public Intellectual Property Resource for Agriculture
PPP	Public–Private Partnership
R&D	Research and development

SACIDS	Southern African Centre for Infectious Diseases and Surveillance
SANBio	Southern African Biosciences Network
SEDA	Small Enterprise Development Agency (South Africa)
SEOBI	SEDA essential oils business incubator
Sida	Swedish International Development Cooperation Agency
SMEs	Small and medium enterprises
SSA	Sub-Saharan Africa
TIA	Technology Innovation Agency (South Africa)
TRIPS	Trade Related Agreement on Intellectual Property Rights
UNEP	United Nations Environment Program
USAID	U.S. Agency for International Development
WCED	World Commission on Environment and Development
WEMA	Water-efficient maize for Africa

Introduction: The Role of a Bio-economy

What is a Bio-economy?

A bio-economy can be thought of as a world where biotechnology contributes to a significant share of economic output [Organization for Economic Co-operation and Development (OECD) 2009]. As conceived by the European Union, a bio-economy involves the sustainable production and conversion of biomass, for a range of food, health, fiber, and industrial products and energy, where renewable biomass encompasses any biological material to be used as raw material (Clever Consult BVBA 2010). The bio-economy can play an important role in creating economic growth, and in formulating effective responses to pressing global challenges.

The development of a bio-economy must take place within the larger context of a green economy, defined by the United Nations Environment Program (UNEP) as "a system of economic activities related to the production, distribution, and consumption of goods and services that result in improved human wellbeing over the long term, while not exposing future generations to significant environmental risks and ecological scarcities" (UNEP 2011). As pointed out by UNEP, economic activity currently consumes more biomass than the Earth produces on a sustainable basis. Hence, increased use of biomass needs to go hand in hand with increased production of biomass.

At the center of much debate concerning the bio-economy and the green economy is the need to find alternatives to conventional fossil fuels. There is a tendency to consider a green economy primarily in terms of energy efficiency and clean energy technologies, with emphasis on biofuels in particular. However, the perils of widespread ecosystem degradation also need to be taken into account (Barbier 2011).

To be sustainable, the development of a bio-economy should focus not only on new uses of biomass, but also on increased biomass production and on its more efficient use, including use of waste products. This puts the need for increased agricultural productivity and better agricultural processing on center stage. The potential of a bio-economy to contribute to the United Nations Millennium Development Goals (MDGs) should not be underestimated. MDG 1 aims to eradicate extreme poverty and hunger by 2015, while MDG 7 aims to ensure environmental sustainability, and MDG 8 addresses the development of a global partnership for development.

What Do We Mean by Innovation in General, and Bio-innovation in Particular?

Innovation has been defined as the introduction of a new idea, product, or process to a user or user-group, and in the context of this chapter refers particularly to the transfer and application of knowledge, research and development, and information in science and engineering (OECD 2009). Innovation does not necessarily involve cutting-edge technology development, and having a smart idea is not enough on its own. Most technology failures occur at the implementation stage, so it is important to develop effective processes to support the complete innovation chain. Knowledge innovation — the science of creating and moving ideas into implementation to benefit an organization, nation, or society as a whole, is perhaps more important than technology innovation on its own.

Particularly in developing countries, innovation tends to be incremental, informal, and mostly below the technology frontier (OECD 2009). Innovation can involve the adoption, refinement, and modification of existing technologies (Brach 2010), but in whatever form it occurs, its importance in economic development should not be underestimated. It has been proposed that the accumulation, absorption, adaptation, production, and transfer of knowledge are at the center of successful development (Stiglitz 2011).

Bio-innovation is all about the innovative application of biotechnology to create products or services. Bio-innovation started by the use of fermentation to expand the range of products that are derived from agricultural crops. This included the development of a range of products derived from agricultural waste. More recently, agricultural biotechnology has seen significant growth, though it is still young with a relatively small stock of innovations (Zilberman and Kim 2011). Many of these innovations show significant potential for African agriculture (Juma 2011), including the more advanced technologies involved in genetic modification, as will be described in more detail, later in this chapter.

In the African context, bio-innovation tends to follow one of three routes. The first involves community-level innovation. This may include, for instance, the application of indigenous knowledge in traditional medicines, the breeding of crops and animals with important characteristics for local conditions, or the use of fermented products as a means of preserving food.

The second route involves technology transfer and adaptation from developed countries; examples include micro-propagation, marker-assisted selection, diagnostics and animal vaccines. These technologies are well-embedded in universities and research institutions in Africa, and are applied to local crops and diseases, and some have even been commercialized.

The third route does not require significant local technological capacity, but involves the implementation of technologies developed elsewhere in the world, in new settings. A good example is the adoption of insect-resistant (Bt) cotton by small-scale farmers in Burkina Faso and elsewhere, after backcrossing into local varieties.

What Do We Understand by Sustainable Development?

As discussed in the previous section, innovation is widely recognized as a key driver of sustainable social and economic development (OECD 2009). To understand this more fully, we also need to understand what we mean by "sustainable development". The World Commission on Environment and Development (WCED) defined it as "development that meets the needs of the present without compromising the ability of future generations to meet their own needs" (WCED 1987). This implies that economic development must ensure that future generations are left no worse off than present generations (Barbier 2011).

Although Africa lags behind much of the rest of the world in economic development, it is clear that both globally as well as in Africa, the current trajectory of economic development is ecologically unsustainable. Globally we are not even meeting the needs of the present, let alone considering the needs of future generations. The planet's biodiversity, by which we mean the rich variety of life forms on Earth, is shrinking, and sustainable development remains an elusive goal. Sub-Saharan Africa (SSA) is one of the regions where natural services are most threatened by human impacts (Millennium Ecosystem Assessment 2005). Crop yields in the region need to increase drastically in order to avoid clearing increasingly more virgin land for agricultural production.

The Role of the Bio-economy in Developing and Developed Countries

Around the world, the development of a bio-economy is very much "work in progress", and strategies to support bio-innovation and the growth of a

Table 5.1 Traditional versus modern bio-economy oriented agriculture

Traditional agricultural development	Modern bio-economy oriented agricultural development
Predominance of agricultural and food security policies	Policy environment integrating natural resources, food and agriculture, energy, and industrial development dimensions
Strong participation and leadership from public institutions as drivers of new technological concepts	Increased use of technologies and products developed in the private sector, public–private partnerships
Agronomic and applied sciences R&D	Horizontal R&D systems ("beyond food"–natural resource use–value chain issues)
Relatively low investment requirements	Higher investment requirements and/or exploitation of technologies developed elsewhere
"Weak" intellectual property systems	"Strong" intellectual property protection systems
Low regulatory intensity	High regulatory intensity (biosafety and consumer protection)
Predominance of bulk marketing and logistical infrastructure, low product differentiation except for quality standards	Increasing importance of value chain integration, product differentiation and standards, and market segmentation issues

Adapted from Trigo (2011)

bio-economy are still emerging. The transition towards a bio-economy requires some structural and policy changes, as shown in Table 5.1.

The EU Bio-economy Strategy, adopted early in 2012, focuses on the potential for conversion of agricultural biomass into food, feed, bio-based products, and bioenergy.

The United States Bio-economy Blueprint was released in April 2012 (United States White House 2012), and focuses on new drugs and diagnostics for improved human health, higher-yielding food crops, biofuels, and bio-based chemical intermediates, to name just a few. Novel technologies such as synthetic biology also feature in this strategy.

In the developing world, South American countries, particularly Brazil and Argentina, are well-positioned to take a lead in developing their bio-economies, as early adopters of biofuel technology and of genetically modified (GM) crops. The Director of Brazil's National Council for Scientific and Technological Development has ambitiously stated that Brazil aims to become the No. 1 player in the global bio-economy (Process Worldwide 2012).

Despite the importance of agricultural biotechnology in any bio-economy strategy, the majority of biotechnology firms globally are focused on the biomedical field (Ernst & Young 2012), rather than on agriculture and environmental sustainability. The OECD Key Biotechnology Indicators Report states that only 12 % of biotechnology firms in OECD countries are active in agriculture (OECD 2011). Agricultural research around the world, but particularly in Africa, has suffered from decades of severe neglect. Yet, with increasing concerns over population growth and associated lack of food security, greenhouse gas emissions, environmental degradation, and climate change, the world is now waking up to the need for more research in these areas.

In Africa, the bio-economy is still emerging as a concept, and there is a need for better political leadership and a new mindset to drive it forward. South Africa is probably the most advanced in this regard, and its Bio-economy Strategy has recently been launched after a long gestation period (South African Department of Science and Technology 2013). Despite the lack of formal policies and leadership in many African countries, aspects of agricultural biotechnology are becoming fairly widely adopted in Africa. As has also been discussed in other chapters of this book, GM crops are a reality in South Africa, Burkina Faso, and Egypt, with Kenya, Uganda, Ghana, Nigeria, Cameroon, and Zimbabwe not far behind. African scientists are using molecular crop breeding techniques, particularly marker-assisted selection, as well as the development of molecular diagnostics (Black et al. 2011; Anthony and Ferroni 2012; Ecuru and Naluyima 2010). Unfortunately, at this stage molecular breeding techniques are not fully integrated into conventional breeding programs in developing countries, and in many SSA countries the introduction of improved crop varieties is hampered by poorly developed seed markets (Anthony and Ferroni 2012).

The best known (and most controversial) contribution of agricultural biotechnology to the bio-economy is the introduction of GM crops, which have had a significant impact on agricultural production in those countries that have adopted them. Transgenic maize, soybean, canola, and cotton have all been widely commercialized, initially with various combinations of insect-resistance and herbicide tolerance, but in the recent past, there has been increasing emphasis on other important traits such as drought tolerance. The United States' revenue in 2010 from GM crops was approximately US $76 billion (United States White House 2012) and adoption rates continue to grow world-wide (James 2011).

At a lower level of technology, other innovations have the potential to contribute significantly to sustainable development in both developed and developing countries. Examples include the use of bacterial and enzyme inoculants for silage and organic fertilizer production, soil inoculants, bio-control products, conversion of agricultural wastes into feed or fuel, and agro-processing for value addition to food.

Some Efforts Towards the Development of a Bio-economy

The stories below are not intended to represent the full spectrum of bio-innovation in Africa, but they are meant to demonstrate some of the diversity of bio-innovation on the continent.

Striga-*Tolerant Cereals*

Striga spp. parasitically infect cereal crops in SSA, commonly causing yield losses of 20–40 %, which can even reach 100 % (Thomson et al. 2010). Across Africa, it is estimated that up to 21 million hectares of farmland are infested with *Striga*

[Association for Strengthening Agricultural Research in East and Central Africa (ASARECA) 2012]. The plant parasite attaches itself to the roots of cereal crops, taking nutrients from the crop in the process. Yields of maize, one of the major cereal crops in the SSA region, are particularly affected.

The chemicals company BASF, as part of their maize-breeding program, developed maize that is resistant to the herbicide Imazapyr. The coating of herbicide-tolerant maize seeds with Imazapyr was conceived as a means of controlling *Striga* infestation, since *Striga* is killed by the herbicide before it can attach itself to the maize root. Herbicide-resistant maize was the first product deployed under the auspices of the African Agricultural Technology Foundation (AATF) (Thomson et al. 2010).

The technology has been deployed in Western Kenya, although adoption levels are still low, and additional promotion and dissemination of the technology is required (Mignouna et al. 2011). Anecdotal evidence suggests that limited availability of seed has constrained uptake of the technology. Additionally, farmers need to be trained in the use of the seed, particularly since farmers who accidentally transfer the herbicide to other crops they plant will inadvertently suffer crop losses. AATF, with support from NGOs, has brought out brochures in local dialects to explain how to handle the seed, and adding to that, disposable gloves are now provided with every packet of seed sold.

Meanwhile, *Striga*-tolerant sorghum has recently been developed through collaboration between Association for Strengthening Agricultural Research in East and Central Africa (ASARECA), the Agricultural Research Corporation (ARC), Sudan, and International Crops Research Institute for the Semi-Arid Tropics (ICRISAT). Taking a different route from maize, the sorghum plants have been developed using marker-assisted selection, with thickened root walls, which makes it impossible for *Striga* to penetrate the root (ASARECA 2012). An alternative approach investigated in Burkina Faso involved inoculation with the fungus *Fusarium oxysporum*, but the marginal rate of return was found to be negative overall (Yonli et al. 2011).

Tissue Culture Bananas

Banana production is decreasing in East Africa due to a variety of pests and diseases, compounded by the degradation of soils. In Central and West Uganda, production is estimated at between 6 and 17 t/ha, compared with the 60 t/ha which is attainable on research stations. The life span of banana plantations has fallen from about 50 years to only 5–10 years in some areas. In Kenya, banana production has also declined over the past decade due to invasions by pests and diseases. The traditional practice of propagation by transplantation of suckers has been a major cause of this problem (Muyanga 2009).

Tissue culture production of banana plantlets offers a means to provide farmers with disease-free planting material. Although Muyanga (2009) found that farmers did not gain any economic benefit from growing tissue culture bananas, this was contested by Kabunga et al. (2012), who showed that after correcting for bias, there was a significant yield gain of 7 %. The same authors found that tissue culture bananas are more responsive to irrigation than conventional bananas, and that improving access to irrigation could boost yield gains by 20 %. Njuguna et al. (2010) were similarly upbeat, stating that the overall direct and indirect economic impact of the tissue culture program in Kenya, in the form of additional income to farm families, amounted to US $94.8 million.

Many African countries, including South Africa, Burundi, Tanzania, Sudan, Ethiopia, Kenya, and Uganda, have significant capacity (human and infrastructural) for plant tissue culture technology (Tushemereirwe et al. 2010). To date, the uptake of banana tissue culture technology in Africa has been relatively low. In Kenya, it is estimated that 5.2 % of banana production is from tissue culture plants (Njuguna et al. 2010). There are a number of constraints, as outlined by Kahangi (2010). These include high costs of production, lack of access to credit by farmers, the need for agricultural inputs, and knowledge of the required agronomic practices.

In Kenya especially, there have been a number of concerted efforts to promote tissue culture banana technology. A project entitled Eastern Province Horticultural Traditional Food Crop Project (EPHTFCP) was conducted between 2001 and 2007, with the goal of increasing the income of smallholder farmers and ensuring food security through increased production of smallholder horticultural crops (Anyango et al. 2010). The project involved the Kenya Agricultural Research Institute (KARI), the Kenya Industrial Research Development Institute (KIRDI), and the Horticultural Crop Development Authority (HCDA) in the introduction and distribution of tissue culture banana planting material, farmer training, market development, and good agricultural practices.

From 2003 to 2007 the Africa Harvest Biotech Foundation International (AHBFI), in partnership with other development agencies, implemented a tissue culture banana project in Central and Eastern Kenya, focused on removing bottlenecks in distribution of plantlets, and improving market access (Njuguna et al. 2010). Africa Harvest is currently working on a United States Agency for International Development (USAID)-supported project, the Kenya Horticultural Competitiveness Project (KHCP), which has distributed over 20,000 banana plantlets to 1,000 farmers during its first year; this was expected to result in over 600 metric tons of produce during the first harvest in August 2012 [Africa Harvest Biotech Foundation International (AHBFI) 2012].

Moving forward, there are opportunities for significant technological innovation to improve banana tissue culture plants. For example, in a project run by the Jomo Kenyatta University in collaboration with the International Institute for Tropical Agriculture (IITA), and funded through the BioInnovate program, technology was developed for inoculation of banana tissue culture seedlings with the endophytic fungus, *Fusarium oxysporum*, for enhanced growth and nematode control (Machungo et al. 2009).

Veterinary Vaccines and Diagnostics

Veterinary vaccines are produced in a number of African countries, including Kenya, South Africa, and Mali. The Pan-African Veterinary Vaccine Centre of the African Union (AU/PANVAC) is based in Ethiopia. The veterinary vaccine producers have of necessity had to innovate, in part because the range of animal diseases is specific to Africa, but also because of the need to ensure the production of good quality vaccines.

In recent years, a major success has been the creation of the Southern African Centre for Infectious Diseases and Surveillance (SACIDS) involving five Southern African countries (South Africa, Tanzania, Zambia, Botswana, Mozambique, Democratic Republic of Congo) together with UK partners, and with funding from the UK Wellcome Trust (Brownlie 2012).

The presence in Africa of a range of diseases that are endemic to the continent has necessitated considerable innovation in vaccine and diagnostics development. Twelve of the 15 known transboundary livestock diseases are to be found in Africa (Brownlie 2012). Some innovative developments have been undertaken together with European partners who are concerned about the spread of disease to that continent. Specific problem diseases include Rift Valley fever, Bluetongue virus, African horse sickness, contagious bovine pleuropneumonia, and haemorrhagic septicaemia.

Currently, the majority of licensed bacterial and viral vaccines in Africa are either live attenuated or inactivated (Lubroth et al. 2007). Nevertheless, technology development is facilitating expression of protective antigens in recombinant vectors, and the development of sub-unit vaccines. Considerable research work is being carried out in Africa, building capacity on the continent to tackle its own important animal health problems. This includes the use of viral vectors as antigen presentation systems (Rutkowska et al. 2011). Particularly innovative work in the area involves the use of plants as a vaccine factory (Rybicki et al. 2012).

Microbial Fermentation Products in South Africa

During the late 1980s, the South African chemical company AECI Ltd. started to move into biotechnology research and development (R&D). The company already had an interest in the agricultural sector through its subsidiaries Kynoch Fertilizers and Kynoch Feeds. After considering all the options, it was decided that the first biotechnology product to be developed should be the amino acid lysine, which is added to maize-based poultry feeds around the world to make up a balanced diet, since maize protein is well-known to be low in this essential amino acid. Lysine is a globally traded commodity, with a market of around 1.5 million tons per annum. There are a limited number of producers world-wide, and in the past, the lysine market has been the subject of litigation regarding cartels and price fixing (Connor 2008).

Lysine is produced by bacterial fermentation using the organism *Corynebacterium glutamicum*. Mutant bacterial strains that produce high levels of lysine are grown under controlled fermentation conditions. AECI was not able to access high-level technology for lysine production from elsewhere, and therefore had to develop its own technology, which took some years to mature to the stage where it was ready for commercial implementation.

During the early 1990s, AECI took the decision to build a commercial lysine plant at their Umbogintwini facility, south of Durban. This was a major step involving significant capital outlay, and represented one of the first ventures into large-scale commercial industrial fermentation on the African continent. The plant, which was constructed in 1995 and commissioned in 1996 at a cost of US $70 million, was small in global terms, with a nominal capacity of 11,000 t (at the time, the average plant in Korea had a capacity of 100,000 t). The plant was originally intended to be a demonstration facility, with the intention of upscaling later, since economies of scale could not be achieved on a small plant. Nevertheless, the level of technological innovation that was involved in getting the plant running efficiently was considerable, and involved the integration of engineering, process, and biological know-how.

During 1997–1998, AECI underwent significant restructuring, resulting in the disbanding of its R&D capability, and a management buyout of the lysine plant. The newly formed company SA Bioproducts continued to produce lysine; but in an increasingly commoditized market, the company also diversified into other products such as the amino acids threonine and isoleucine. In 2009, the company was acquired by the Canadian company Lallemand Inc., and has since started yeast production.

While the history of SA Bioproducts demonstrates the difficulty of operating in global markets, at the same time it must be recognized that very significant capacity has been developed, and that South Africa now has considerable expertise in the production of fermentation-based products. Further ventures into large-scale fermentation plants would necessarily involve levels of capital outlay that may be beyond the scope of African investors.

Agro-processing

Agro-processing offers considerable opportunities for Africa to add value to its agricultural produce, and it is ideally suited for the development of small and medium enterprises (SMEs). Agro-processing reduces post-harvest losses, extends shelf life, and improves the quality and safety of foods [Food and Agriculture Organization/United Nations Industrial Development Organization (FAO/UNIDO) 2010]. Processing of citrus, pineapple, and oil palm are good examples. However, access to agricultural produce needs to be linked to appropriate technology and finance, and there is considerable potential for growth of this sector. The examples below are intended to give just an idea of some developing trends in agroprocessing bio-innovation.

In Ghana, small companies such as Kona Agroprocessing Ltd., have been assisted by the Grassroots Business Fund to process cashew nuts for sale to export markets. Kona is part of a wave of new cashew-processing firms in West Africa that aim to bring economic benefit at the local level. Meanwhile, in Nigeria, a major investment of US $150 million is planned to build a 600 t agroprocessing complex focusing on cassava value addition (NXP Online 2012).

Agro-processing industries are fairly well developed in East Africa. For example, in several East African countries the Mt. Meru Group is expanding the opportunities for locally grown oilseeds by processing sunflower and soybean to produce edible oils. They have recently partnered with the Rwandan Government and the Clinton Foundation to establish a new soybean processing factory in Rwanda. Meanwhile, the Kenyan Government is planning to establish four pilot SME agro-processing parks (Kenya Vision 2030 2011). A number of new opportunities are being developed through the Swedish funded Bio-resources Innovations Network for Eastern Africa Development (Bio-Innovate) Program, which is supporting a range of exciting multidisciplinary and multi-country projects (Bio-Innovate 2012).

In South Africa, the Council for Scientific and Industrial Research (CSIR) has supported the growth of a number of small agroprocessing SMEs. The CSIR has a benefit-sharing agreement with traditional healers, who supplied CSIR scientists with information about the traditional use of an indigenous plant, *Lippia javanica*, as a mosquito repellent. The CSIR then developed mosquito-repellent candles based on an extract of essential oils from the plant, and have transferred the technology for cultivation and processing of the plants to community-owned businesses in rural areas. A community-based candle-manufacturing factory has been established, and mosquito-repellent candles are now on sale in various shops in South Africa (Maharaj et al. 2008).

The CSIR has also assisted other communities to establish agroprocessing businesses. In the small community of Dysselsdorp in the Cape, a business has been established to extract liquorice from the roots of the naturalized plant, *Glycyrrhiza glabra*. The liquorice blocks are now exported to markets around the world. Other essential oils businesses involving plants such as the rose geranium have also been established in previously impoverished communities.

Challenges Facing Bio-entrepreneurs in Africa

Access to Finance, Markets, and Business Advisory Services

One of the first challenges facing would-be bio-entrepreneurs in Africa is the development of a viable business plan. The fact that the private sector is relatively poorly developed means that mentors who have real experience in business development are in short supply. Most innovations arise from work in universities or research institutes, yet the scientists concerned have little or no knowledge of how to translate their work into a product or process, whether for commercial gain, for community empowerment, or for global public good.

In most cases, for a business strategy to be viable, there must be a sufficiently large market to justify the cost of development. It is usually easiest to introduce a new product into a local market where it can prove its worth, before it is rolled out elsewhere. This is not always easy in Africa, where there is limited purchasing power. Transport costs and inefficiencies can hamper access to markets in the developed world, although some industries, such as the horticulture industry in Kenya, have managed to overcome these hurdles.

Although access to seed funding and venture capital is a problem for bio-entrepreneurs around the world, the problem is particularly acute in Africa. Although venture capital is available, the high-risk profile and prolonged time required to market, associated with most biotechnology projects tend to make this an unattractive sector for investors. In some cases, African governments are stepping up to at least partially fill the funding gap, through organizations such as the Technology Innovation Agency, in South Africa. Some donor organizations are providing funding to venture capital companies such as African Agricultural Capital, which is dedicated to agriculture, though not specifically to biotechnology (Gatsby Foundation 2011).

Likewise the African Innovation Foundation, founded by three Swiss-based Angolans, seeks to plug a gap in the innovation chain, but it is not sector-specific.

A relatively new initiative, 3ADI, the African Agribusiness and Agro-industries Development Initiative, involving the African Union, various United Nations Agencies, the International Fund for Agricultural Development (IFAD), and the African Development Bank, is pursuing mechanisms to stimulate the development of agriculture and agroprocessing in selected African countries (FAO/UNIDO 2010), but this initiative will not cover the full range of opportunities offered by biotechnology in Africa.

Time and Cost of Product Development

Most product development initiatives in Africa are undertaken in the face of a shortage of critical mass, in terms of human resources, physical facilities, and finance. As a result, product development takes too long, and may in fact never reach completion. It has been shown that innovation speed (defined as the time from initial concept to product commercialization) is positively correlated with product success (Kessler and Bierly 2002). So long as African innovation lags behind the rest of the world, it will be difficult for the continent to be truly competitive. The other components of a successful innovation strategy are quality (including product design, standardization, and packaging), and efficiency (producing maximum innovative output with the available resources); these two components are also negatively affected by lack of skilled resources allocated to product development.

Intellectual Property Hurdles

For SSA countries, intellectual property rights (IPRs) are both a burden and an opportunity (Blakeney and Mengistie 2011). Article 7 of the Trade Related Agreement on Intellectual Property Rights (TRIPS) states that the protection and enforcement of IPRs "should contribute to the promotion of technological innovation and to the transfer and dissemination of technology, to the mutual advantage of producers and users of technological knowledge and in a manner conducive to social and economic welfare" (World Trade Organization 1994). Although many have criticized the TRIPS agreement, conversely, the weak protection of IPRs in many African countries may act as a hindrance to innovation. Companies may be reluctant to transfer technology to Africa if they feel their rights to the technology will not be respected. A critical issue for developing countries is therefore how IPRs might help or hinder them in gaining access to technologies that are required for their development (Blakeney and Mengistie 2011).

Yet in African countries with stronger IPR protection, such as South Africa, where there have been many patents filed, other problems arise. The fragmented ownership of IPRs across multiple public and private sector owners produces situations where few single institutions can provide a complete set of IP rights to ensure freedom-to-operate (FTO) with any given technology (Chi-Ham et al. 2012), resulting in particular difficulties for public institutions wishing to innovate in the biotechnology sector. Additional problems faced by public sector institutions are lack of understanding of intellectual property (IP) issues by developing-country researchers and technology managers, as well as the high cost of patenting; it may not be financially viable to protect technology in all the relevant countries. Chi-Ham et al. (2012) outline the role of the Public Intellectual Property Resource for Agriculture (PIPRA) in attempting to overcome some of these problems.

In the case of GM crops, companies engaging in technology transfer for philanthropic reasons and/or through public–private partnerships are becoming concerned about the potential impact of the recently adopted Kuala Lumpur Supplementary Protocol with relation to the Cartagena Protocol on Biosafety, in dealing with liability and redress. The terms of the Protocol will be embedded in the national laws of African countries, raising concerns that in donating intellectual property for the benefit of developing countries, companies could be taking on additional risks of legal liabilities for the technology.

At the same time, there are concerns that IP regimes are inappropriate to protect the rights of holders of traditional knowledge (Prasad et al. 2012). It is necessary to ensure that benefits accrue for communities that hold indigenous knowledge, even though this might not be protected through patents, but unless there is adequate IP protection, no company will be willing to invest resources in developing a commercial product using information derived from traditional knowledge.

Legislative Hurdles

The introduction of GM crops has encountered numerous legislative hurdles not only in Africa but also elsewhere in the world. Many African countries have opted for very strict legislation that imposes an impossibly high regulatory burden. As pointed out by Farre et al. (2011), in the short-to-medium term some important GM crops for Africa are unlikely to be commercialized because politicians, supported by propaganda from activists, the public, and the media, pressure the regulators to increase the regulatory burden.

Regulatory hurdles also exist for a range of non-GM technologies, and it is important that appropriate regulatory structures are put in place at the same time that technologies are being developed. For example, the introduction of *Striga*-resistant maize has been delayed in some East African countries because of stringent requirements to register the herbicide Imazapyr. Another example is the need for regulations to facilitate the introduction of biofuels that would lay down appropriate standards and specifications.

Funding for Research and Development

The lack of government support for research and development, including biotechnology R&D, is a major deterrent to the development of the biotechnology sector, and results in researchers becoming reliant on the agendas of international donor organizations (Morris 2011). Short-term, non-sustainable funding also means that researchers are often unable to complete projects before the funding runs out. In an era of strong competition for funds, scientists find themselves forced to over-promise on deliverables, resulting in eventual disappointment, and disillusionment on the side of the funders.

The funding constraints likewise impact on availability of equipment. While many laboratories are chronically under-equipped, with aging and poorly maintained instruments, the situation is even worse when it comes to scaling up for commercial implementation. The majority of funders do not consider funding of pilot scale equipment as one of their priorities.

Technical Skills and Know-How

Brach (2010) suggested that two-thirds of variation in economic development can be explained by a country's technological readiness. Countries with low technological readiness are hampered in their ability to innovate by adopting and modifying existing technologies; citing Egypt as an example, the author points out that countries which can access technologies, but have limited ability to use and apply them, seem to be unable to systematically transform innovative activities into knowledge and products.

The important role of universities in providing technical know-how is highlighted in a Nigerian study (Onyeka 2011). Not only in Nigeria, but in other African countries, there are many factors hampering both the development of cutting-edge skills in biotechnology, and the translation of bioscience or biotechnology from an academic endeavor into commercialized outcomes.

Examples of Mechanisms to Overcome the Challenges

Bio-incubators

Biotechnology incubators have become recognized around the world as a standard way to support nascent business development and to bridge the gap from research to commercialization. They provide biotechnology entrepreneurs with access to shared equipment and facilities, assistance with business planning and business development, and a supportive environment for product development. The majority of business incubators are sponsored by public agencies, government, or academic institutions (Tonukari 2008).

The African Union document "Freedom to Innovate" (Juma and Serageldin 2007) advocates the creation of "Local Innovation Areas", which would serve as focal points for innovation activity, and would be effective in helping to incubate new business start-ups while giving a boost to companies on their way to becoming more established. Tonukari (2008) likewise stresses the importance of bio-incubators if biotechnology is to take off in Africa.

While at this stage, the shortage of government investment in biotechnology is hampering the establishment of bio-incubators in Africa, in the United States there are over 100 bio-incubators. The only bio-incubator-related activity in the African continent appears to be in South Africa. The South African Department of Trade and Industry, through the Small Enterprise Development Agency (SEDA), provides funding to over 30 business incubators, yet only three of these relate to the biotechnology sector: eGoliBio (Bio and Life Sciences), Makfura Makhura Incubator (Biofuels), and the SEDA essential oils business incubator (SEOBI) (plant-derived essential oils), [Small Enterprise Development Agency (SEDA) 2011]. Unfortunately, government funding to the incubators is at sub-critical levels, severely constraining the development of the sector. The new Technology Innovation Agency, funded through the South African Department of Science and Technology, has not yet lived up to its early promise, but will hopefully soon fill some of the gaps in funding and support for new business development.

Chakma et al. (2010) provides a good overview of the lessons learned from the development of an earlier South African bio-incubator, Acorn Technologies. They point out that even where there are severe funding constraints, incubators can still deliver value by operating as a virtual organization with little physical infrastructure, focusing on entrepreneurship training and networking, while maintaining strict selection criteria for incubatees.

Public–Private Partnerships

Traditionally, the public and private sectors have attempted to provide solutions independently from each other (Ferroni and Castle 2011). But in recent years, Public–Private Partnerships (PPPs) have emerged as a popular type of development collaboration. PPPs offer the prospect of overcoming the limitations of both the public and the private sector: the public sector has limited ability to market its research outputs, while the private sector will not engage in activities where the market is insufficient to ensure profitability. Ferroni and Castle (2011) cite specific examples of PPPs involving the Syngenta Foundation for Sustainable Agriculture, such as the development of the semi-dwarf *Eragrostis tef*, the staple food of Ethiopia; rust-resistant wheat; micronutrient-enhanced crops through the CGIAR HarvestPlus Challenge Program, including vitamin A-enhanced sweet potato (Uganda and Mozambique), maize (Zambia), and cassava (Nigeria); and iron-rich pearl millet (India) and bean (Rwanda).

The African Agricultural Technology Foundation (AATF), headquartered in Kenya, specifically has a mandate to facilitate and promote PPPs. One of their flagship projects is the Water Efficient Maize for Africa (WEMA) project, a partnership between Monsanto, BASF, the CGIAR International Maize and Wheat Improvement Center (CIMMYT), and the national agricultural research systems in participating countries. Funding is provided by the Bill and Melinda Gates Foundation and the Howard G. Buffet Foundation. This complex project demonstrates the potential of multi-partner PPPs to deliver value above and beyond what could be possible through any one organization.

Regional African Collaborations

There have been, and are, a number of international collaborations for agricultural research in Africa. Of particular note are the well-funded CGIAR centers, many of which have activities and research institutes in Africa. Yet the international nature of these organizations means that they tend to be isolated from national research activities. It is only in recent years that regional networks have emerged, driven by the regions themselves (Roseboom 2011). Such collaborations include CORAF/WECARD (the West and Central African Council for Agricultural Research and Development), and ASARECA (the Association for Strengthening Agricultural Research in Eastern and Central Africa). At an Africa-wide level, FARA (the Forum for Agricultural Research in Africa) promotes collaboration in agricultural research across the continent, and runs various projects that support networking, including an internet portal eRAILS for communication of agricultural innovations.

Under the auspices of the New Partnership for Africa's Development (NEPAD), additional regional collaborations have been established including BecA (Biosciences East and Central Africa), SANBio (Southern African Biosciences Network), etc. These initiatives serve to promote research capacity, support collaborative research projects, and provide access to equipment and facilities at a regional level.

Donor Funded Programs

As described above, donor funding is most often provided for specific projects such as the WEMA project. Yet in some cases, sustained broad-based program funding over a number of years has alleviated some of the problems associated with lack of funding continuity, and allowed true capacity development and innovation to occur.

A good example of this is the BioInnovate (Bio-resources Innovations Network for Eastern Africa Development) program, funded by the Swedish International Development Cooperation Agency (Sida). This developed from the earlier Bio-EARN program (Eastern Africa Regional Program and Research Network for Biotechnology, Biosafety and Biotechnology Policy Development). Through three phases of Bio-EARN, starting in 1999, capacity was built in biotechnology, biosafety, and biotechnology policy development in Ethiopia, Kenya, Tanzania, and Uganda. The follow-up BioInnovate program, which was started in 2010, has included additional countries (Rwanda and Burundi) and has built on the partnerships that were created through Bio-EARN to fund collaborative projects that improve crop productivity and resilience to climate change in small-scale farming systems, and to increase the efficiency of the agroprocessing industry to add value to local bio-resources in a sustainable manner. The long-term commitment, over considerably more than a decade, is remarkable.

Another example of long-term commitment is the funding provided by USAID to ABSP (Agricultural Biotechnology Support Project). The first phase of ABSP ran for 12 years from 1991 to 2003. Although its activities extended beyond Africa, it provided significant support to transgenic biotechnology projects in Kenya, South Africa, and Egypt, particularly focusing on potatoes resistant to the tuber moth, and sweet potatoes with resistance to Feathery Mottle Virus. At the same time, support was provided for the development of appropriate policies and legislation to deal with these technologies (Brink 2003). The second phase of ABSP (ABSPII) is still running, but the only project in Africa receiving support is the development of transgenic bananas in Uganda (Cornell University 2012).

Develop IP to a Certain Point then Out-License and Earn Income Through Royalties

It should be recognized that African countries do not necessarily have the resources to take their intellectual property all the way to the stage of getting a product on the market. Where there are severe regulatory hurdles, it may be necessary to out-license technology with the aim of receiving royalties in return. Good examples would be the development of pharmaceutical products from medicinal plants, or the commercialization of GM crops. For example, the CSIR in South Africa, in consultation with the San people, where the knowledge originated, licensed the IP for the appetite suppressant derived from the Hoodia plant to a multinational company. Although this product has never reached the market, interim payments have been made, providing at least some return for the technology development.

Finding Novel Ways to Increase Access to High End Equipment

The NEPAD regional bioscience networks are playing an important role in facilitating access to expensive equipment. Particularly, the BecA (Biosciences East and Central Africa) hub based at ILRI in Kenya, and the SANBio (Southern African Biosciences network) hub based at the CSIR in Pretoria, play a key enabling role, not only in providing access to equipment and facilities, but also through the associated expertise and training that they can provide. This type of approach demonstrates that even though African countries may not have the same level of resources as the developed world, with some ingenuity and cooperation it is still possible to achieve results.

As the bioscience world becomes more and more data-driven through the rise of genomics and other associated "-omics" techniques, it is also becoming apparent that competitive success may arise less from the ability to generate data oneself, but more from the ability to analyze and interpret data. This is where bioinformatics comes into play, and where African countries may in the future be able to compete with the rest of the world on a more level playing field supported by rapidly improving broadband internet connectivity and bandwidth availability.

What Africa Needs to Develop a Bio-economy

The development of an agricultural bio-economy in Africa involves not only increased agricultural production, but also the development of competitive export-based agro-industries.

African countries must focus on areas where they have a real competitive advantage. Competitive advantage can be achieved in a variety of ways: through access to unique biodiversity including indigenous food plants, through access to traditional knowledge, through climatic advantages, through the availability of land suitable for agriculture, or simply by having a wide range of agricultural, environmental, and health problems that call for the development of products to address them.

African countries already rank low on the global competitiveness index. A recent effort to adjust this index to measure "sustainable competitiveness", one of the hallmarks of a bio-economy, resulted in the surveyed African countries dropping even further down the list (World Economic Forum 2012). Investment climate reforms are essential; the cost of doing business in Africa is 20–40 % above that for other developing regions. At the same time, a major investment in skills development is needed. The bio-economy requires biotechnologists who not only understand their own discipline but have all the skills at their fingertips to participate in the bio-economy. Surveys report that African tertiary graduates are weak in problem-solving, business understanding, computer use, and communication skills (World Bank survey as reported by Page 2012).

A rapid transition to a bio-economy is essential; time is not on the side of the continent. This may involve elements of disruptive change as frequently observed in the development and adoption of new technologies, but will also require managed change to ensure that supportive policies are in place. The rapid adoption of cell phone technology in Africa provides an excellent example of how a new technology can take off if the conditions are right. The lack of pre-existing investment in fixed line infrastructure meant that Africa could adopt the new technology without the systemic inertia of more developed countries.

Strong leadership by Africa's politicians is required to remove generic roadblocks such as customs barriers to regional trade, to establish systems to ensure that farmers have access to modern agricultural inputs, to promote African-based integrated value chains, and above all to promote a change of mindset regarding the importance of science, technology, and innovation.

An integrated approach to the development of the bio-innovation value chain is required. The current South African situation demonstrates the problems that arise from fragmentation and inefficiency. Efforts to address the so-called "innovation chasm" between academic research and commercialization in South Africa resulted in the formation of the Biotechnology Innovation Centres (BICs), which were subsequently absorbed into the newly created Technology Innovation Agency (TIA). Unfortunately, lack of strong leadership, inefficient implementation, lack of expertise, and turf wars between government entities have resulted in a loss of morale, and continuing gaps in the innovation chain.

A regional approach to the development of a bio-economy could ensure that lessons learned are shared widely, that synergies are identified and built on, and that unnecessary duplication of facilities and instrumentation is avoided. Collaboration occurs between scientists in the region at a project level, but African governments need to provide much stronger support to regional initiatives. Even the regional bioscience networks established under NEPAD are showing signs of strain due to inadequate high-level support from within the continent. Unfortunately, despite strong-sounding statements at the level of the African Union, there seems to be considerable inertia and lack of accountability when it comes to implementation; this has to change.

Conclusions and Recommendations

A number of opportunities have been touched upon in the sections above, including improved crops, animal vaccines, and medicinal products derived from the continent's biodiversity. Other opportunities include biofuels, biological fertilizers, biological control agents, plant and animal diagnostics, biological waste treatment, nutraceuticals, and cosmeceuticals.

There are already many products in the development pipeline around the continent, too numerous to list here. In the short run, value addition to crops and crop by-products through agro-processing is a significant focus and a major opportunity. The investigation of medicinal plants is also a major research focus, but here the pathway to commercialization can be more complex.

The development of diagnostics is receiving increasing attention from the research community, but there is scope for more attention to be paid to diagnostics of plant and animal diseases that are of particular relevance to Africa. There is a real opportunity here to partner with scientists in the north, to produce low-cost diagnostic kits.

Agricultural yields in Africa continue to be low, due to lack of agricultural inputs and poor quality planting materials. While biotechnology cannot compensate for problems of this nature, the development of improved planting material with resistance to pests and diseases, and tolerance to climatic extremes, particularly drought, is a priority that is deservedly receiving much attention. In combination with biological fertilizers and biological control agents, Africa has a real opportunity to develop sustainable agriculture.

Africa has a higher proportion of young people than any other continent; around 70 % of the continent's population is under 30 years of age. The bio-economy offers huge opportunities for job creation and entrepreneurship amongst this population, where unemployment levels are at alarmingly high levels. African governments must recognize the potential for unrest due to disaffected youth, and must rapidly move to stimulate and support bio-innovation and the bio-economy.

By adopting all the opportunities that biotechnology has to offer, Africa has an opportunity to leapfrog ahead of the rest of the world. A focused and sustained effort over the next few years is needed to demonstrate just what can be achieved.

References

Africa Harvest Biotech Foundation International (AHBFI) (2012) With over 600 mt of harvest expected in August, Africa Harvest shifts focus to value-addition of TC bananas. Retrieved July 2013, from http://africaharvest.org/articles.php?id=98

Anthony VM, Ferroni M (2012) Agricultural biotechnology and smallholder farmers in developing countries. Curr Opin Biotechnol 23:278–285. doi:10.1016/j.copbio.2011.11.020

Anyango JJ, Wambugu FM, Nkanya J, Kyalo G, Onyango C (2010) Introduction of tissue culture banana technologies and their impact on producer wellbeing. Acta Hortic 879:767–772

Association for Strengthening Agricultural Research in East and Central Africa (ASARECA) (2012) Retrieved 9 Oct 2012, from http://www.asareca.org/content/hope-african-sorghum-farmers-striga-resistant-variety-released-0

Barbier E (2011) The policy challenges for green economy and sustainable economic development. Nat Resour Forum 35:233–245

Bio-Innovate (2012) Retrieved 9 Oct 2012, from http://bioinnovate-africa.org/projects

Black R, Fava F, Mattei N, Robert V, Seal S, Verdier V (2011) Case studies on the use of biotechnologies and on biosafety provisions in four African countries. J Biotechnol 156:370–381. doi:10.1016/j.jbiotec.2011.06.036

Blakeney M, Mengistie G (2011) Intellectual property and economic development in sub-Saharan Africa. J World Intellect Prop 14(3/4):238–264. doi:10.1111/j.1747-1796.2011.00417.x

Brach J (2010) Technological readiness in the Middle East and North Africa — implications for Egypt (GIGA working paper 155). GIGA research programme 3: socio-economic challenges in the context of globalisation. Retrieved 3 Oct 2012, from http://ssrn.com/abstract=1743389

Brink J (2003) Agricultural biotechnology support project final technical report 1991–2003. Retrieved 12 Nov 2012, from http://pdf.usaid.gov/pdf_docs/PDACD491.pdf

Brownlie J (2012) A foresight vision for infectious diseases in Africa. Onderstepoort J Vet Res 79(2):454–459, http://dx.doi.org/10.4102/ojvr.v79i2.459

Chakma J, Masum H, Singer PA (2010) Can incubators work in Africa? Acorn Technologies and the entrepreneur-centric model. BMC Int Health Hum Rights 10(1):S7, Retrieved 12 Nov 2012, from http://www.biomedcentral.com/1472-698X/10/S1/S7

Chi-Ham CL, Boettiger S, Figueroa-Balderas R, Bird S, Geoola JN, Zamora P, Alandete-Saez M, Bennett AB (2012) An intellectual property sharing initiative in agricultural biotechnology: development of broadly accessible technologies for plant transformation. Plant Biotechnol J 10:501–510

Clever Consult BVBA (2010) The knowledge based bio-economy (KBBE) in Europe: achievements and challenges. Report produced for the Belgian Presidency and the European Commission. Retrieved 12 Nov 2012, from http://www.bio-economy.net/reports/files/KBBE_2020_BE_presidency.pdf

Connor JM (2008) Evolution of the global lysine industry, 1960–2000 (Purdue J Paper 16283). Retrieved 12 Nov 2012, from http://ssrn.com/abstract=1188512

Cornell University (2012) Agricultural biotechnology support project II. Retrieved 12 Nov 2012, from http://www.absp2.cornell.edu/

Ecuru J, Naluyima H (2010) Biotechnology developments in Uganda and associated challenges. Afr Crop Sci J 18(4):133–139

Ernst & Young (2012) Beyond borders (Global biotechnology report 2012). Retrieved 12 Nov 2012, from http://www.ey.com/GL/en/Industries/Life-Sciences/Beyond-borders—global-biotechnology-report-2012

Farre G, Twyman RM, Zhu C, Capell T, Christou P (2011) Nutritionally enhanced crops and food security: scientific achievements versus political expediency. Curr Opin Biotechnol 22: 245–251. doi:10.1016/j.copbio.2010.11.002

Ferroni M, Castle P (2011) Public-private partnerships and sustainable agricultural development. Sustainability 3:1064–1073. doi:10.3390/su3071064

Food and Agriculture Organization/United Nations Industrial Development Organization (FAO/UNIDO) (2010) African agribusiness and agro-industries development initiative (3ADI): a programme framework. Retrieved 12 Nov 2012, from http://www.hlcd-3a.org/data_all/PDF_en/a3ADI_progFram1.pdf

Gatsby Foundation (2011) African agricultural capital. Retrieved 9 Oct 2012, from http://www.gatsby.org.uk/~/media/Files/Africa/Gatsby%20Africa%20AAC%20Two%20Page%20Summary%202011.ashx

James C (2011) Global status of commercialized biotech/GM crops: 2011. ISAAA Brief No. 43. ISAAA, Ithaca

Juma C (2011) The new harvest: agricultural innovation in Africa. Oxford University Press, New York

Juma C, Serageldin I (2007) Freedom to innovate: biotechnology in Africa's development. Report for African Union, New Partnership for Africa's Development, Addis Ababa/Pretoria

Kabunga NS, Dubois T, Qaim M (2012) Yield effects of tissue culture bananas in Kenya: accounting for selection bias and the role of complementary inputs. J Agr Econ 63(2):444–464. doi:10.1111/j.1477-9552.2012.00337.x

Kahangi EM (2010) The potential of tissue culture banana (*Musa* spp.) technology in Africa and the anticipated limitations and constraints. Acta Hortic 879:281–288

Kenya Vision 2030 (2011) Retrieved 12 Nov 2012, from http://www.vision2030.go.ke/index.php/pillars/project/Economic/39

Kessler EH, Bierly PE III (2002) Is faster really better? An empirical test of the implications of innovation speed. IEEE T Eng Manage 49(1):2–12

Lubroth J, Rweyemamu MM, Viljoen G, Diallo A, Dungu B, Amanfu W (2007) Veterinary vaccines and their use in developing countries. Rev Sci Tech 26(1):179–201

Machungo C, Losenge T, Kahangi E, Coyne D, Dubois T, Kimenju J (2009) Effect of endophytic *Fusarium oxysporum* on growth of tissue-cultured banana plants. Afr J Hortic Sci 2:160–167

Maharaj VJ, Fouche G, Senabe J, Nthambeleni R, Kotze F (2008) Agro-processing opportunities identified through a novel mosquito repellent from a medicinal plant. Retrieved 12 Nov 2012, from http://researchspace.csir.co.za/dspace/bitstream/10204/2744/1/Maharaj_P_2008.pdf

Mignouna DB, Manyong VM, Rusike J, Mutabazi KDS, Senkondo EM (2011) Determinants of adopting Imazapyr-resistant maize technologies and its impact on household income in western Kenya. AgBioForum 14(3):158–163

Millennium Ecosystem Assessment (2005) Retrieved 12 Nov 2012, from http://www.millenniumassessment.org

Morris EJ (2011) Modern biotechnology — potential contribution and challenges for sustainable food production in sub-Saharan Africa. Sustainability 3:809–822. doi:10.3390/su3060809

Muyanga M (2009) Smallholder adoption and economic impacts of tissue culture banana in Kenya. Afr J Biotechnol 8(23):6548–6555

Njuguna MM, Wambugu FM, Acharya SS, Mackey MA (2010) Socio-economic impact of tissue culture banana (*Musa* spp.) in Kenya through the whole value chain approach. Acta Hortic 879:77–86

NXP Online (2012) Retrieved 12 Nov 2012, from http://www.nxponline.com/2012/03/150m-agro-processing-complex-up-in.html

Onyeka CJ (2011) Biotechnology commercialization in universities of developing countries: a review of the University of Ibadan, Nigeria: research and regulation. J Commer Biotechnol 17:293–300
Organization for Economic Co-operation and Development (OECD) (2009) The bio-economy to 2030: designing a policy agenda. Retrieved 12 Nov 2012, from http://www.oecd.org/futures/thebio-economyto2030designingapolicyagenda.htm
Organization for Economic Co-operation and Development (OECD) (2011) Key biotechnology indicators. Retrieved 12 Nov 2012, from http://www.oecd.org/science/innovationinsciencetechnologyandindustry/49303992.pdf
Page J (2012) Can Africa industrialise? J Afr Econ 21(2):ii86–ii125. doi:10.1093/jae/ejr045
Prasad R, Bagde US, Varma A (2012) An overview of intellectual property rights in relation to agricultural biotechnology. Afr J Biotechnol 11(73):13746–13752
Process Worldwide (2012) Retrieved 12 Nov 2012, from http://www.process-worldwide.com/management/markets_industries/articles/345478/
Roseboom J (2011) Supranational collaboration in agricultural research in sub-Saharan Africa (ASTI/IFPRI – FARA conference working paper 5). Conference. Agricultural R&D: investing in Africa's Future. Retrieved 12 Nov 2012, from http://www.ifpri.org/sites/default/files/publications/roseboom.pdf
Rutkowska DA, Meyer QC, Maree F, Vosloo W, Fick W, Huismans H (2011) The use of soluble African horse sickness viral protein 7 as an antigen delivery and presentation system. Virus Res 156:35–48, http://dx.doi.org/10.1016/j.virusres.2010.12.015
Rybicki EP, Chikwamba R, Koch M, Rhodes JI, Groenewald JH (2012) Plant-made therapeutics: an emerging platform in South Africa. Biotechnol Adv 30:449–459
Small Enterprise Development Agency (SEDA) (2011) SEDA technology programme annual review 2010/11. Retrieved 12 Nov 2012, from http://www.seda.org.za/Publications/Publications/Seda%20Annual%20Report%202010-2011.pdf
South African Department of Science and Technology (2013) The Bio-Economy Strategy. ISBN 978-1-9966-06-04. 44pp. http://www.dst.gov.za/images/ska/Bioeconomy%20Strategy.pdf
Stiglitz JE (2011) Rethinking development economics. World Bank Res Obs 26(2):230–236. doi:10.1093/wbro/lkr011
Thomson JAT, Shepherd D, Mignouna HD (2010) Developments in agricultural biotechnology in sub-Saharan Africa. AgBioForum 13(4):314–319
Tonukari NJ (2008) Africa needs biotechnology incubators. Biotechnology 7(4):803–807
Trigo EJ (2011) The bio-economy in Latin America and the Caribbean: towards a socioeconomic research agenda. LAC regional IAAE inter-conference symposium on the bio-economy, CIAT, Cali, 19–20 Sept 2011
Tushemereirwe WK, Byabachwezi M, Njuguna J, Nankinga CK, Nowakunda K, Ngambeki D, Namaganda JM, Ssemwanga J (2010) The role of national centers of excellence in banana (*Musa* spp.) research and development in Eastern, Central, and Southern Africa. Acta Hortic 879:289–293
United Nation Environment Program (UNEP) (2011) Towards a green economy: pathways to sustainable development and poverty eradication. Retrieved 12 Nov 2012, from http://www.unep.org/greeneconomy/greeneconomyreport/tabid/29846/default.aspx
United States White House (2012) National bio-economy blueprint. Retrieved 12 Nov 2012, from http://www.whitehouse.gov/sites/default/files/microsites/ostp/national_bio-economy_blueprint_april_2012.pdf
World Commission on Environment and Development (WCED) (1987) Our common future. Retrieved 12 Nov 2012, from http://www.un-documents.net/wced-ocf.htm
World Economic Forum (2012) The global competitiveness report 2012–2013. Retrieved 12 Nov 2012, from http://www3.weforum.org/docs/WEF_GlobalCompetitivenessReport_2012-13.pdf
World Trade Organization (1994) Agreement on trade-related aspects of intellectual property rights. Retrieved 12 Nov 2012, from http://www.wto.org/english/docs_e/legal_e/27-trips.pdf

Yonli D, Traoré H, Van Mourik TA, Hess DE, Sérémé P, Sankara P (2011) Integrated control of *Striga hermonthica* (Del.) Benth. in Burkina Faso through host plant resistance, biocontrol and fertilizers. Int J Biol Chem Sci 5(5):1860–1870

Zilberman D, Kim E (2011) The lessons of fermentation for the new bio-economy. AgBioForum 14(3):97–103

Chapter 6
Biotechnology Success Stories by the Consultative Group on International Agriculture Research (CGIAR) System

Melaku Gedil, Leena Tripathi, Marc Ghislain, Morag Ferguson, Marie-Noelle Ndjiondjop, Lava Kumar, Bodo Raatz, Luis Augusto Becerra Lopez-Lavalle, Ranjana Bhattacharjee, Kassa Semagn, and Jean-Marcel Ribaut

Abstract The CGIAR (Consultative Groups for International Agricultural Research, www.cgiar.org) deploys agricultural biotechnology innovations to improve crops' and livestock's productivity and quality in Africa. CGIAR centers have played a pivotal role in kick-starting agbiotech among NARS partners in Africa through various contributions towards building human resource and infrastructure as well as providing access to genomic resources of major African crops and developing biotech varieties and associated biosafety regulatory systems. Genomic resources such as molecular markers, genetic linkage maps, transcriptome, annotated genome sequences, which are extremely valuable for molecular breeding were limited or non-existent for a large number of African staple crops until recently. Modern breeding schemes aimed at accelerating genetic gain, such as genome selection (GS), marker-assisted recurrent selection (MARS), and marker-assisted back-crossing (MABC) are underway for many African crops. In an effort to deploy molecular breeding by NARS, genotyping services and

M. Gedil (✉) • L. Kumar • R. Bhattacharjee
International Institute of Tropical Agriculture (IITA), Ibadan, Nigeria
e-mail: m.gedil@cgiar.org

L. Tripathi • M. Ferguson
International Institute of Tropical Agriculture (IITA), Nairobi, Kenya

M. Ghislain
International Potato Center, Regional Office, Nairobi, Kenya

M.-N. Ndjiondjop
Africa Rice Center, Cotonou, Benin

B. Raatz • L.A.B. Lopez-Lavalle
International Center for Tropical Agriculture (CIAT), Cali, Colombia

K. Semagn
International Maize and Wheat Improvement Center (CIMMYT), Nairobi, Kenya

J.-M. Ribaut
Generation Challenge Program (GCP), Mexico City, Mexico

F. Wambugu and D. Kamanga (eds.), *Biotechnology in Africa*,
Science Policy Reports 7, DOI 10.1007/978-3-319-04001-1_6,

Web-based Integrated Breeding Platform (IBP, https://www.integratedbreeding.net/) providing crop information and analytical tools to help design and conduct marker-assisted breeding experiments have been developed. With regard to plant protection, CGIAR centers have played a significant role in the development and application of molecular tools for the characterization, detection and diagnosis of disease causing agents through development of simple and accurate tools and procedures. Harnessing advances in biotechnology tools and increasing availability of genomes of pathogens and pests helped CGIAR centers and partners to address the complexity of pathogen diversity, germplasm evaluation, and monitoring of mycotoxins in food and feed samples as well as efficient assays for simultaneous detection of multiple pathogens. Genetic engineering has been applied to improvement of priority traits where conventional non-GM approaches have little promise. Several biotech products are now in the pipeline with anticipated release time in the coming few years. Examples include transgenic banana plants that have exhibited strong resistance to banana *Xanthomonas* wilt (BXW), Nematode resistant plantain, virus and weevil resistant sweetpotato, late blight resistant potato, and cassava brown streak disease (CBSD) resistant cassava. Application of additional biotechnological tools such as doubled haploid technologies, next-generation sequencing based applications, and genome editing technologies are poised to further accelerate the impact of biotechnology in enhancing agricultural productivity in Africa.

Keywords Biotechnology • CGIAR • Marker-assisted breeding • Genetic engineering • Molecular diagnostics • Genetic resources

Abbreviations

AATF	African Agricultural Technology Foundation
AfricaRice	Africa Rice Center
ALS	Angular leaf spot
BCMNV	Bean common mosaic necrotic virus
BCMV	Bean common mosaic virus
BGYMV	Bean golden yellow mosaic virus
BXW	Bacterial Xanthomonas wilt
CAPS	Cleaved amplified polymorphic sequence
CBB	Common bacterial blight
CBSD	Cassava brown streak disease
CGIAR	Consultative Group on International Agriculture Research
CIAT	International Center for Tropical Agriculture
CIMMYT	International Maize and Wheat Improvement Center
CIP	International Potato Centre (Peru)
CMD	Cassava mosaic virus
DTMA	Drought-tolerant maize for Africa
ELISA	Enzyme-linked immunosorbent assay

GBS	Genotyping by sequencing
GCP	Generation Challenge Program (Mexico)
GSS	Genotyping support services
GWS	Genome-wide selection
HRAP	Hypersensitive response-assisting protein
IBP	Integrated breeding platform
ICRISAT	International Crops Research Institute for the Semi-Arid Tropics
IITA	International Institute of Tropical Agriculture
KARI	Kenya Agricultural Research Institute
LAMP	Loop-mediated isothermal amplification
MAB	Marker assisted breeding
MABC	Marker-assisted backcrossing
MARS	Marker-assisted recurrent selection
MAS	Marker-assisted selection
NARL	National Agricultural Research Laboratory (Uganda)
NARO	National Agricultural Research Organisation (Uganda)
NARS	National agricultural research system
NERICA	New Rice for Africa
NGS	Next generation sequencing
PCR	Polymerase chain reaction
PFLP	Plant ferredoxin-like protein
QTL	Quantitative trait loci
R4D	Research for development
RAPD	Random amplified polymorphic DNA
RMC	Red mottled advanced lines for the Caribbean
RYMV	Rice yellow mottle virus
SCAR	Sequence-characterized amplified region
SiRNA	Small interfering RNA
SNP	Single nucleotide polymorphism
SSA	Sub Saharan Africa
SSR	Simple sequence repeat
TALEN	Transcription activator-like effector nucleases
UCBSV	Ugandan cassava brown streak virus
WEMA	Water-efficient maize for Africa

Introduction

Agricultural biotechnology is a collection of several scientific techniques such as genetic engineering, molecular breeding, molecular diagnostics, tissue culture, and vaccines, in order to improve crop and livestock productivity and the marketability of derived products. This chapter provides an insight into the activities and achievements of the CGIAR (Consultative Groups for International Agricultural Research, www.cgiar.org) in deploying agricultural biotechnology in Africa. The

CGIAR coordinates the research activities of 15 independent, non-profit agricultural research centers. In this chapter, we highlight selected success stories of the CGIAR centers from the 1960s to date. Various bottlenecks have impeded the significant impact of biotechnology on crop improvement in developing countries (Ribaut and Ragot 2007). Limited human resources and inadequate field infrastructure remain major challenges and, until recently, breeders in those countries had, among other issues, limited access to genomic resources for their target crops and few predictive molecular markers for use in breeding programs (Ribaut et al. 2010). The introduction and application of agricultural biotechnology in Africa has been greatly facilitated by several CGIAR centers, which have helped in the creation of centers of excellence for modern DNA technologies to researchers in Africa. With increased investment in biotechnology over the past 2 decades, CGIAR centers continue to be important contributors to building capacity around agric-biotechnology through the training of national agricultural research systems (NARS) partners.

Development of Molecular Tools to Enhance Crop Improvement

Resource-poor African farmers largely depend on orphan crops for their diet and income. The benefit of orphan crops lies in their adaptation to extreme environmental conditions such as drought/heat, infertile soil, and other biotic and abiotic stresses. Biotechnology has tremendous potential to improve the productivity and nutritional quality of these crops (Tadele and Assefa 2012; Varshney et al. 2012). The CGIAR system generated many species-specific simple sequence repeat (SSR) markers for their mostly 'orphan' mandate crops, such as groundnut (Ferguson et al. 2004), pigeon pea (Odeny et al. 2007), banana (Buhariwalla et al. 2005), cassava (Mba et al. 2001), potato (Ghislain et al. 2009), and sweet potato (Tumwegamire et al. 2011). Several genome sequences have been partially uncovered in the last few years (cassava in 2009, potato in 2011, banana in 2012, yam projected in 2013), while others are soon to be completed. These milestones have added to the existing genomic resources. In this respect, sweet potato and yam are orphan crops with only partial transcriptome analyzed (Schafleitner et al. 2010; Narina et al. 2011). Single nucleotide polymorphisms (SNPs) have rapidly become the most widely used markers as a result of their cost-effectiveness and abundance in the genome. With relative improvement in the cost of assays and skills, technologies have gradually permeated African research institutions. Recent reviews on the status of genomic resources and application in crop improvement have been provided for cassava (Ferguson et al. 2011) and yam (Bhattacharjee et al. 2011). The Generation Challenge Program of the CGIAR (GCP) helped overcome some of the myriad constraints, through the development of genotyping resources, called genotyping support services (GSS) based on SSR, SNP, and diversity array technology markers for a broad range of crops, with a particular focus on less-studied crops, also referred to as 'orphan crops' (Varshney et al. 2010). These resources are now readily accessible, including high-throughput marker technologies at a level that now makes molecular breeding a reality for those important staple crops in developing countries (https://www.integratedbreeding.net/snp-marker-conversion).

Genetic Resources Management

Collectively, the CGIAR centers conserve in their gene banks about 650,000 samples of crops, forage, and agroforestry genetic resources as public goods. However, most of these collections are large in size, impeding their proper use in crop improvement programs. Hence, there is a need to develop strategies which use innovative technologies for characterizing genetic diversity. In a global effort galvanized by the GCP, several major and underutilized species have been characterized using different types of molecular markers to assess genetic diversity. Core collections have been developed in several crops, including the crops maintained in CGIAR genebanks such as groundnuts, pearl millet, yams, cassava, and cowpea. More recently, mini-core collections (Upadhyaya et al. 2009) have been developed in different crops, offering immense opportunities to identify new sources of variation for use in crop improvement. The advent of high-throughput technologies such as next generation sequencing (NGS) provides a better understanding, and adds value to collections stored in genebanks (Kilian and Graner 2012). Efforts are underway to use NGS techniques such as genotyping by sequencing (GBS) to characterize core collections of cassava and yams. AfricaRice Genebank contains about 20,000 accessions of *O. sativa* and 2,500 accessions of *O. glaberrima* and wild species. Molecular characterization of several subsets of the genebank was performed using molecular markers (Semon et al. 2005).

Marker-Assisted Breeding

Marker assisted breeding (MAB) is the process of using the results of DNA tests to assist in the selection of individuals to become the parents in the next generation of a genetic improvement program. The choice among various methods of MAB depends on the complexity of the trait and a prior knowledge of the genes or segments of chromosomes (known as quantitative trait loci or QTL). Molecular markers will help breeders in one or more of the following ways: facilitate conventional breeding, improve selection efficiency, reduce cost for developing new varieties, and/or quality control (control line purity and genetic identity). There are several ongoing projects in Africa that utilize molecular markers for developing improved varieties for drought tolerance, low nitrogen tolerance, insect resistance,

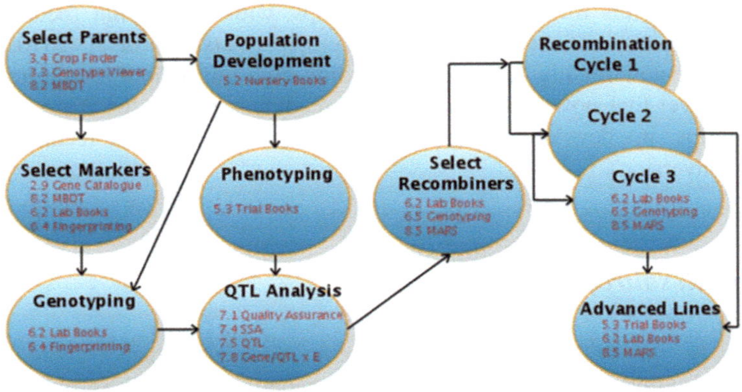

Fig. 6.1 Breeding workflow for a marker assisted recurrent selection experiment (Adapted from: Ribaut et al. 2012)

disease resistance, etc. Contrary to the belief that molecular breeding requires cost and sophisticated facilities, breeders in Africa only need a very small lab for extracting and quantifying DNA samples before shipping to regional genotyping service providers (e.g., BecA) or service providers in the US and Europe. Capacity building in MAB pipeline is therefore a vital and priority need. The GCP is coordinating the development and deployment of a sustainable Web-based integrated breeding platform (IBP, https://www.integratedbreeding.net/) as a one-stop shop for information, analytical tools, and related services to help design and conduct MAB experiments in the most efficient way. Such a platform will enable breeding programs in developing countries to accelerate variety development using marker technologies for different breeding purposes: major genes or transgene introgression via marker-assisted backcrossing (MABC), gene pyramiding via marker-assisted selection (MAS), marker-assisted recurrent selection (MARS, Fig. 6.1) and, in a not too distant future, genome-wide selection (GWS) (Delannay et al 2012). With these introductory remarks, we present selected examples of the successful use of biotechnological intervention in the improvement of African crops involving CGIAR centers.

Cassava Improvement

Breeding for superior varieties in clonally propagated crops such as cassava, yam, banana, and potato is particularly challenging, due to the biology of these crops. In a longstanding partnership and a concerted effort to harness state-of-the-art molecular technologies, IITA and CIAT and multiple public and private partners have played a leading role in enriching genomic resources for cassava, making it one of the most resource-rich crops in the region. Recent reviews (Ferguson et al. 2011; Okogbenin et al. 2007) provide an insight into available genomic resources, and describe the role biotechnology played in improving key traits such as yield, disease resistance, drought tolerance, and quality traits. The recent rapid accumulation of cassava resources, such as genome sequence, transcriptome, numerous SSR and SNP markers and assays, triggered several mega projects aimed at leveraging genomics for an efficient and effective cassava breeding strategy. The most recent project is the innovative genome selection breeding approach termed the 'Nextgen cassava' project (http://nextgencassava.org).

Potato Improvement

A significant number of qualitative genes have been mapped on the potato genetic linkage map, and diagnostic markers for MAS exist (Mori et al. 2011; Li et al. 2013). A few commercial breeding programs are using these for potato virus Y, nematode *Globodera rostochiensis*, and *G. pallida* (Ortega and Lopez-Vizcon 2012). However, many important traits are governed by multiple QTL with small effects and unknown epistatic effects. International Potato Centre (CIP) is starting to use markers associated with a resistance gene to the virus PVY, but these are still not being used effectively for MAS. This situation is likely to change with technologies providing high throughput makers at very low cost, and the development of computer software handling SNP data from polyploid crops.

Sweet Potato Improvement

Though an important African crop, sweet potato has been long neglected by breeders. It was hoped that DNA markers would improve selection efficiency in the crop. Inheritance of resistance to virus disease and DNA markers were characterized a decade ago, but none have yet proven to be useful for selection (Mwanga et al. 2002). Other important breeding traits are dry-matter, starch, and β-carotene content, for which DNA markers are now available (Cervantes-Flores et al. 2011). Similarly to potato, these polygenic traits have proven difficult to benefit from with MAB, due to the genetic nature of the crop. However, NGS and new bio-informatics tools may finally bring the resolution needed when assigning markers to phenotypic value, so that these will be effective in predicting parental value as well as selecting genotypes. Significant investment will, however, be needed from public funding organizations, as the crop has so far attracted little interest from the private sector.

Maize Improvement

In maize breeding programs, the maintenance of inbred line genetic purity and confirmation of the genetic identity of genotypes are also important quality control functions. Molecular markers can be used for various purposes, including quality control genotyping, germplasm characterization, selection of parental combinations, and MAB. Using MABC, Ribaut and Ragot (2007) introgressed five QTLs associated with yield components and flowering in maize from a donor parent in a drought-susceptible recurrent parent. The best MABC progeny outperformed the recurrent parent by two to four times under severe drought conditions, with no yield reduction under optimal conditions. Currently, CIMMYT, in collaboration with the national agricultural research systems (NARS) from 14 countries in Africa, the International Institute of Tropical Agriculture (IITA), the African Agricultural Technology Foundation (AATF), Monsanto Company, and several regional and national seed companies in Africa, is involved in the drought-tolerant maize for Africa (DTMA) and water-efficient maize for Africa (WEMA) projects that aim at developing drought-tolerant maize for Sub Saharan Africa (SSA) using conventional breeding, MARS, and/or transgenic technology. In the latest ambitious project named SeeD of Discovery, CIMMYT has unleashed the power of biotechology to mine useful traits in the broadest reserve of genetic variability of maize in the world by genotyping the entire collection to rapidly characterize and create new breeding material.

Rice Improvement

In 1992, the Africa Rice Center (AfricaRice) and its partners started to work on interspecific hybridization in an attempt to combine the useful traits of *O. sativa* and *O. glaberrima* (Jones et al. 1997). The back-crossing of *O. glaberrima* to *O. sativa* was coupled with another culture to overcome the reproductive barrier between cultivated interspecific crosses to gain access to valuable traits, resulting in the development of New Rice for Africa (NERICA) lines. NERICA lines have been tested in 31 SSA countries and more than 1,000,000 ha are now under upland NERICA production. Recently, a collaborative project between IRD and AfricaRice has identified more than 500,000 *O. glaberrima*-specific SNPs. AfricaRice is using *O. glaberrima* as a vital reservoir of useful genes for discovery and breeding.

Molecular tools have been successfully used to genetically characterize rice yellow mottle virus (RYMV)-resistant genes discovered in some accessions (Albar et al. 2006; Thiémélé et al. 2010). The fine mapping and gene cloning allowed the easy transfer of the Rymv1 resistance gene into popular varieties susceptible to the virus through MAB.

Due to poor water management in most of the lowland areas of SSA, including irrigated fields, rice plants are often affected by submergence at various intensities and durations. Most existing rice cultivars are seriously damaged if they are completely submerged for more than 3 days. However, a few tolerant cultivars can withstand complete submergence for 10–14 days, such as FR13A from which a major QTL, named SUB1, has been identified on chromosome 9 (Xu and Mackill 1996). Development of submergence-tolerant rice varieties using MAS is well under way at AfricaRice (Iftekharuddaula et al. 2011).

Marker Assisted Selection in Bean Breeding

A molecular marker for resistance to bean golden yellow mosaic virus (BGYMV) is one of the earliest and most used examples of markers use in breeding of common bean. A SCAR marker named SR2 based on a co-dominant RAPD marker (Urrea et al. 1996) was identified for the *bgm-1* resistance gene. Various lines have been released originating from material selected with this marker, e.g., the Caribbean (RMC) series of red mottled advanced lines (Blair et al. 2006). Liebenberg et al. (2006) used three co-dominant SCAR markers to introgress rust resistance into cultivars tracing *Ur-13* for pyramiding with other resistance loci in breeding lines. Resistance genes *Ur-3, Ur-5*, and *Ur-11* of Mesoamerican origin are being deployed by MAS in bean cultivars for East Africa due to their high effectiveness against the Andean races of rust pathogens (Miklas et al. 2006). Tryphone et al. (2012) described the introgression of CBB resistance in the locally adapted Tanzanian cultivar Kablanketi. Gene stacking for several foliar diseases was facilitated by using molecular markers, as demonstrated in an ongoing study at CIAT-Uganda to pyramid four genes conferring resistance to angular leaf spot (ALS), anthracnose, Pythium, and bean common mosaic and necrotic viruses (BCMV and BCMNV) to improve the durability of resistance.

The development of high throughput SNP assays such as a 5 k genotyping chip (BeanCap, Hyten et al. 2010), genotyping by sequencing (GBS) and Fluidigm platforms, together with the availability of common bean draft genome sequence (http://www.phytozome.com), allows positioning of molecular markers on the map aiding genomic studies. As these platforms are still cost-prohibitive for regular analysis of large breeding populations, other methods that cost-effectively genotype small numbers of SNPs are also utilized. For instance, a CAPS marker has been developed for the very important *bc-3* gene for BCMV resistance based on an SNP published by Naderpour et al. (2010), allowing genotyping on agarose gels. For the same SNP, an assay has been developed using an inexpensive gel-free system using real-time PCR (Wang et al. 2005). More than 1,500 SNPs (in cooperation with BeanCAP) have been made available through the commercial LGC/Kbioscience SNP platform (http://www.lgcgenomics.com/genotyping), facilitated by GCP, for genotyping hundreds of bean lines. This gives researchers in Africa the ability to adopt the benefit of the genomics era to improve their breeding programs without major investments in instrumentation.

Biotechnology in Pathogen Diagnostics

CGIAR centers have played a significant role in the development and application of molecular tools for the characterization, detection, and diagnosis of disease-causing agents. The focus of most molecular diagnostics programs has generally been on developing tools and technologies for better understanding of disease ecology, diagnosis, and monitoring of biological systems to prevent the trans-boundary spread of pests and pathogens and their negative impacts on plant health, crop production, and quality of the produce. There has been greater emphasis on the development of simple and accurate tools and procedures for rapid diagnosis of pathogens and pests of food and horticultural crops in sub-Saharan Africa.

Initial applications of molecular diagnostic tools targeted viral pathogens. IITA, ICRISAT and CIP specialize in the production of poly- and monoclonal antibodies against viruses, and the development of enzyme-linked immunosorbent assay (ELISA)-based detection tools (Agindotan et al. 2003). ELISA-based diagnostics were established for over 30 economically important viruses (Devi et al. 1999), which were used to formulate low-cost serological diagnostics kits. Antibodies were also produced against nonprotein targets such as mycotoxins.

Since the 1990s, polymerase chain reaction (PCR)-based tools have been established as the basis for the handling of a range of viruses, due to their high sensitivity and rapid turn around time. This has enabled the discrimination of strains and closely related species, earlier not possible with ELISA-based assays. PCR-based assays have proven particularly convenient for several targets such as groundnut rosette virus (Naidu et al. 1998) and unculturable pathogens like phytoplasma (Kumar et al 2011a).

Advances in biotechnology tools, and the increasing availability of genomes of pathogens and pests, have created new opportunities for CGIAR-led R4D approaches to address the complexity of pathogen diversity (Abang et al. 2006; Legg and Fauquet 2004; Kumar et al. 2011b), germplasm evaluation (Tripathi et al. 2008), and monitoring of mycotoxins in food and feed samples (Waliyar et al. 2008). Further improvement in the form of a multiplex PCR assay developed for the simultaneous detection of all the begomovirus species involved in cassava mosaic disease etiology reduced the cost and time involved in disease diagnosis (Alabi et al. 2008). Of the various applications, the most significant use of molecular diagnostic tools has been in epidemiological investigations mapping pathogen distribution and spread (Legg et al. 2011; Kumar et al. 2011b).

Genetic Engineering

Direct gene transfer technologies that result in crop production are referred to as transgenics, and offer the opportunity to add important traits to an existing well-adapted and -adopted variety when the gene(s) are known and have a dominant effect. It also has the distinct advantage over other biotechnologies that it adds traits for which the crops' own genetic resources do not have the corresponding genes (i.e., resistance to insect pests using proteins from bacteria, Bt crops). However, its drawback is the high cost of regulation due to perceived risks on human and environmental health. The CGIAR has, therefore, prioritized its transgenics products for traits of significant incidence on food security and nutrition in Africa. A few examples follow.

Fig. 6.2 Disease evaluation of transgenic banana plants in screen house, (**a**) Non-transgenic inoculated plants showing complete wilting after 60 days of post inoculation of *Xanthomonas campestris* pv. *musacearum* (*Xcm*), (**b**) Transgenic inoculated banana plants showing no symptoms after 60 days of post inoculation of *Xcm*. (**c**) Resistance to late blight of a transgenic event bearing the RB gene from *Solanum bulbocastanum* from the potato variety Desiree: on the *left* an untransformed plant, on the *right* the transgenic event #70, both inoculated with the isolate Pox67 of *Phytophthora infestans*

Banana with Resistance to Xanthomonas Wilt

Bacterial wilt caused by *Xanthomonas campestris* pv. *musacearum* (BXW) threatens banana production and the livelihoods of smallholder growers. Given the rapid spread and devastation of BXW across the continent, and the absence of resistant cultivars, genetic transformation through the use of modern biotechnology tools offers an effective, fast, safe, and viable way to develop resistant varieties (Tripathi et al. 2009). Transgenic banana plants that have exhibited strong resistance to BXW in the laboratory, and screen house tests have been developed recently (Fig. 6.2) (Tripathi et al. 2010; Namukwaya et al. 2012). Sixty-five of the most resistant lines were planted in a confined field trial at the National Agricultural Research Laboratory (NARL), Kawanda, Uganda, for further evaluation. Twelve transgenic lines have shown absolute resistance to BXW. The HRAP and PFLP proteins used for transformation are not listed as potential allergens, and the risk of gene flow from the plant under consideration to another crop species is not an issue in edible banana. Nevertheless, the transgenic lines will be tested for food and environmental safety in keeping with biosafety regulations.

Nematode-Resistant Plantain

IITA, in partnership with the University of Leeds, the UK has developed a transgenic plantain with a high degree of nematode resistance using maize cystatin and synthetic repellent genes. Hundreds of independent transgenic lines of the plantain cultivar 'Gonja manjaya' were generated from embryogenic cell suspensions (Roderick et al. 2012). The lines expressing the transgenes were evaluated, among which many lines provided significant resistance to *R. similis*, showing resistance levels of 70–84 %. The promising transgenic lines showing high resistance to nematodes were planted in a confined field in Uganda in December 2012 for further evaluation. There is no concern about the safety of cystatin-based transgenic work, as it has been frequently undertaken and is well-established for rice cystatins.

Virus and Weevil Resistant Sweet Potato

Sweet potato productivity is hampered by the incidence of virus diseases and weevil damage. CIP and NARO, Uganda have meshed their expertise to develop weevil resistance in sweet potato using Bt technologies. Similarly, the Donald Danforth Plant Science center has, in partnership with NARO and CIP, developed transgenic strains that could potentially withstand viruses and other diseases. Eventually, weevil and virus disease resistance will be combined and allow farmers to realize a quantum productivity gain while reducing crop losses during the dry season, when weevil-related losses usually occur (Mwanga et al. 2011).

Late Blight Resistant Potato

Irish potato has an important role in Africa as a subsistence crop in highlands and a cash crop for many smallholders. Two diseases are responsible for significant yield losses, going up to 10–15 % annually: late blight and bacterial wilt. CIP in partnership with advanced research institutions and NARO, Uganda and KARI, Kenya, are developing late blight resistant potato varieties (Fig. 6.2). Genes from wild species of potato are stacked to confer durable resistance to this disease. A high level of resistance has already been observed through confined field trials. The research on bacterial wilt engineering is still in its early stages, but has the potential to do away with a major constraint on potato production in Africa, a feat which is not presently achievable by other means.

CBSD-Resistant Cassava

Cassava brown streak disease (CBSD), has emerged as the biggest threat to cassava cultivation in East Africa. As known sources of resistance for the disease are difficult to introduce into farmer-preferred cultivars by conventional methods, integration of resistance traits via transgenics holds significant potential in the tackling of this disease. Of the available transgenic approaches to control plant viral diseases, RNA silencing is a very promising strategy that has been successfully employed to control viral diseases. Cassava plants were engineered to generate small interfering RNAs (siRNAs) from the UCBSV coat protein (ΔCP) sequence (Ogwok et al. 2012).

Prospects and Future Direction

The affordability and accessibility of next-generation sequencing technology is bound to transform the field of applied genomics, thereby accelerating the ability to decipher the genetic factors behind agriculturally important traits. The technology is readily available for researchers in developing countries, as there is no need for capital investment (outsourcing is the most cost-effective alternative). However, it should be noted that full exploitation of these technologies demands a high level of computational competency (bioinformatics and data management), which is still a constraint in Africa's institutions (Gedil 2009). With the cost of genotyping out of the way, the focus is on accurate phenotyping. In SSA, the diversity of agro-ecological zones allows for the setting up of high-throughput precision phenotyping facilities, including disease hotspots, drought/heat tolerance screening sites, low fertility soils, and aluminum toxicity screening sites among others.

Doubled Haploids in Cassava, Banana, and Potato Inbred Lines

Producing haploid plants containing chromosomes from only one parent creates lines that are immediately inbred (homozygous) after they are reconverted into fertile diploids ("doubled haploids"). A novel strategy for producing haploids based on centromere engineering (Ravi and Chan 2010) will be used to create haploid cassava and banana by expressing an altered transgenic protein called CENH3. The resultant transgenic plants will be crossed to wild type clones at both CIAT and IITA, and haploid individuals recovered. Potato breeding has already used the ability of certain diploid accessions to reduce the ploidy of tetraploid varieties to di-haploids, but these retain the parental hetereozygosity. However, near homozygous potatoes have been developed in diploid potato after the introgression of the S-locus inhibitor *Sli* gene from *Solanum chacoense* (Lindhout et al. 2011). At CIP, we will explore this new approach by exploiting the genetic diversity of Andean diploid potatoes in order to develop hybrid seed with heterosis.

Synthetic Seeds for Clonal Propagation of Disease-Free Cassava

Most diseases such as CMD spread quickly and easily from one cropping cycle to the next, due to the lack of certified, clean propagation material. Cassava is almost exclusively propagated clonally through cuttings obtained from infected plants. The adoption of newly released varieties is slowed due to the lack of sufficient healthy planting material. Sexual seed (true seed), which usually impedes the transmission of diseases, is often not used for cassava multiplication since several cultivars do not flower or, if they do, the resultant offspring is not genetically identical to the mother plant due to the heterozygous nature of the crop. There are unexplored, potential methods to massively propagate cassava using non-sexual, synthetic seeds (synseeds). Synseeds derive from totipotent plant cells, tissues, and organs capable of generating complete clonal plants. Somatic embryos (SEs) are made of totipotent cells. They are routinely produced in cassava, mostly for genetic modification purposes, although not yet available for massive propagation. This technique offers cassava farmers an alternate method of multiplying disease-free cassava.

Efficient Pathogen Diagnostics Tools

More recently, the power of modern high-throughput DNA sequencers is being harnessed to understand evolution and variability of viruses (Kreuze et al. 2013). A new initiative at IITA is focused on application of high-throughput proteomics to identify protein biomarkers to rapidly identify variation in vectoring potential of aphid and whitefly vector populations. Using technology based on siRNA sequencing, CIP and its partners are determining the 'virome' of sweet potato throughout Africa. Virus content, strain diversity, and geographic distribution will be a major practical output which will be useful for adjusting phytosanitary policies and containment measures across the continent. Platforms for the sensitive detection of multiple viruses such as microarrays may be developed as practical and efficient solutions for national diagnostic laboratories. ClonDiag tube arrays for potato and sweet potato, as well as loop-mediated isothermal amplification (LAMP) are being tested for their sensitivity and ease of use. Numerous challenges still loom on the road to bringing these technologies to the field, but the benefits are worth the effort. In addition to technology development, efforts are being made to transfer technology, products, and skills to stakeholders in national research and extension services through collaborative activities and organization of training courses.

Editing Genomes and Genes

Genetic engineering will deliver transgenic products with significant benefits to resource-poor farmers. However, the cost of regulation and controversies associated with this technology will limit its applicability to traits not achievable by other means. New methods to modify genomes have been developed to reduce the cost of regulation, such as intragenics and cisgenics and others using engineered DNA-binding proteins (Carlson et al. 2012). The resulting genetic modification can range from single nucleotide changes up to full gene insertion. These technologies, referred to as gene or genome editing, have numerous potential applications, from altering gene regulation to shuffling alleles. Since the end product is equivalent to the product of random mutagenesis or plant breeding, it has been portrayed as a non-transgenic technology. However, it is not yet clear whether an international consensus will emerge on how to regulate these genetic modifications. At CIP, we will investigate the use of TALEN to deactivate as well as introduce resistance genes in the potato.

Conclusion

Genomic resources such as genetic linkage maps, transcriptome, and annotated genome sequences are extremely valuable in the optimization of the future applications of molecular breeding. Genomic resources were, however, limited or non-existent for a large number of African staple crops until recently. Various CGIAR projects have endeavored to circumvent this constraint through concerted, multi-partner efforts for most of the key crops. The outcome was enhanced biotechnological capacity of the national agricultural research programs in several countries through training of scientists and building of infrastructure. The development and implementation of the Integrated Breeding Platform (IBP) by GCP allowed access and utilization of virtual/cyber tools and information exchange among African scientists, and facilitated research in advanced institutions. This achievement laid a strong foundation for the deployment of molecular breeding in Africa. Several young scientists and graduate students are already taking advantage of these tools.

On the other hand, for some major African crops such as maize, rice, cassava, and potato, a wealth of genomic and plant breeding resources are available, including genome sequences, transcriptomes, and dense genetic linkage maps. However, the use of this information and related resources is still very limited in Africa, especially in the national breeding programs. Crop improvement requires long-term commitment in building both human capacities and infrastructure, as well as sustained funding to hire qualified personnel and purchase and maintain the necessary equipment and reagents. In many African countries, availability of sufficiently trained breeders in conventional and molecular plant breeding, as well as appropriate infrastructure (including phenotyping infrastructure) supported by adequate financial resources, appear to be the key constraints hindering effective and productive molecular breeding programs (Delannay et al. 2012). Also, due to the lack of incentives to maintain qualified and competent staff, there is a high staff turnover rate in most NARS. The logistics of reliably shipping perishable reagents is also often an obstacle because of long procurement and clearance procedures. To address some of these issues, different initiatives have been undertaken, such as the individual training and multi-year training of young students and NARS researchers by the GCP, the AGRA plant breeding training program, and the BecA platform. However, to be sustainable, such efforts should be supported by appropriate and conducive national policies and investments in agricultural research.

Biotech crops resulting from the application of transgenesis are also another yet unrealized opportunity to solve production constraints and nutrition deficiencies. Since its first commercial success nearly 20 years ago, the adoption of this technology has been erratic due to economic and ideological conflicts. The safety of this procedure, to both humanity and the environment, has been demonstrated over and over without a single case of proven damage. In spite of this record of safety and success, biotech crops are slow to be embraced in Africa. However, those crops engineered for traits not achievable by other means are likely to be adopted soon. Virus-resistant cassava, bacterial wilt- and nematode-resistant banana, weevil- and virus-resistant sweet potato, late blight-resistant potato — these are just a few examples of the future of biotech crops in Africa. Eventually, once these demonstrate benefits to resource-poor farmers and create food security, the adoption of such crops will be accelerated.

This chapter provides a brief account of the deployment of biotechnology research by CGIAR centers for improving agricultural productivity in selected African staple food crops. Cognizant of the rapid advances in genomics, strategies are being laid out to enhance the adoption and integration of biotechnology, including bioinformatics, into agricultural research geared towards the development of climate-resilient, quality, and productive agricultural products. CGIAR and NARS scientists must continue to drive the change of crop improvement methodologies towards higher resolution and efficiency by accessing all tools derived from biotechnology. Stewardship of the products of genomics and biotechnology will continue to be critical, as will investments from donors and CGIAR centers. The fast changing field of biotechnology offers tremendous opportunity for revolutionizing African agriculture.

References

Abang MM, Asiedu R, Hoffmann P, Wolf GA, Mignouna HD, Winter S (2006) Pathogenic and genetic variability among *Colletotrichum gloeosporioides* isolates from different yam hosts in the agroecological zones in Nigeria. J Phytopathol 154:51–61

Agindotan BO, Thottappilly G, Uwaifo A, Winter S (2003) Production of monoclonal and polyclonal antibodies against a Nigerian isolate of banana streak virus. Afr J Biotechnol 2:171–178

Alabi OJ, Kumar PL, Naidu RA (2008) Multiplex PCR method for the detection of African cassava mosaic virus and East African cassava mosaic Cameroon virus in cassava. J Virol Methods 154:111–120

Albar L, Bangratz-Reyser M, Hebrard E, Ndjiondjop M-N, Jones M, Ghesquière A (2006) Mutations in the eIF(iso)4G translation initiation factor confer high resistance to Rice yellow mottle virus. Plant J 47:417–426

Bhattacharjee R, Gedil M, Sartie AM, Otto E, Dumet D, Kikuno H, Kumar PL, Asiedu R (2011) Dioscorea. In: Kole C (ed) Wild crop relatives: genomic and breeding resources. Springer, Heidelberg, pp 71–96

Blair MW, Beaver JS, Nin JC, Prophete E, Singh SP (2006) Registration of PR9745–232 and RMC-3 red-mottled dry bean germplasm lines with resistance to bean golden yellow mosaic virus. Crop Sci 46:1000–1002

Buhariwalla HK, Jarret RL, Jayashree B, Crouch JH, Ortiz R (2005) Isolation and characterization of microsatellite markers from *Musa balbisiana* (Primer Note). Mol Ecol Notes 5:327–330

Carlson DF, Fahrenkrug SC, Hackett PB (2012) Targeting DNA with fingers and TALENs. Mol Ther Nucleic Acids 1(1):e3

Cervantes-Flores JC, Sosinski B, Pecota KV, Mwanga RO, Catignani GL, Truong VD, Watkins RH, Ulmer MR, Yencho GC (2011) Identification of quantitative trait loci for dry-matter, starch, and β-carotene content in sweet potato. Mol Breed 28(2):201–216

Delannay XG, McLaren G, Ribaut JM (2012) Fostering molecular breeding in developing countries. Mol Breed 29:857–873

Devi KT, Mayo MA, Reddy KL, Delfosse P, Reddy G, Reddy SV, Reddy DVR (1999) Production and characterization of monoclonal antibodies for aflatoxin B1. Lett Appl Microbiol 29:284–288

Ferguson ME, Burow MD, Schultz S, Bramel P, Paterson AH, Kresovich S, Mitchell S (2004) Microsatellite identification and characterization in peanut (*A. hypogaea* L.). Theor Appl Genet 108(6):1064–1070

Ferguson M, Rabbi IY, Kim D, Gedil M, Lopez-Lavalle LA, Okogbenin E (2011) Molecular markers and their application to cassava breeding: past, present and future. Trop Plant Biol 5(1):95–109

Gedil M (2009) Tailoring bioinformatics for the genetic improvement of orphan crops. Afr Technol Dev Forum 6:34–43

Ghislain M, Núnez J, del Rosario HM, Pignataro J, Guzman F, Bonierbale M, Spooner DM (2009) Robust and highly informative microsatellite-based genetic identity kit for potato. Mol Breed 23(3):377–388

Hyten DL, Song Q, Fickus EW, Quigley CV, Lim JS, Choi IY, Hwang EY, Pastor-Corrales M, Cregan PB (2010) High-throughput SNP discovery and assay development in common bean. BMC Genomics 11:475

Iftekharuddaula KM, Newaz MA, Salam MA, Ahmed HU, Mahbub MAA, Septiningsih EM, Collard BCY, Sanchez DL, Pamplona AM, Mackill DJ (2011) Rapid and high-precision marker assisted backcrossing to introgress the SUB1 QTL into BR11, the rainfed lowland rice mega variety of Bangladesh. Euphytica 178:83–97

Jones MP, Audebert A, Mande S, Aluko K (1997) Characterization and utilization of *Oryza glaberrima* Steud. in upland rice breeding. In: Jones MP, Dingkuhn M, Johnson DE, Fagade SO (eds) Interspecific hybridization: progress and prospects. Proceedings of the workshop: Africa/Asia Joint Research on Interspecific Hybridization between the Africa and Asian Rice Species (*O. glaberrima* and *O. sativa*). West Africa Rice Development Association, Bouaké, pp 43–59

Kilian B, Graner A (2012) NGS technologies for analyzing germplasm diversity in genebanks. Brief Funct Genomics 2:38–50

Kreuze JF, Fuentes S, Muller M, Gutierrez DL, Boonham N, Fei Z (2013) Viral diagnostics: from continental viromes to field level diagnostic methods. In: Book of abstracts, the 12th international plant virus epidemiology symposium, IITA, Ibadan, p 60

Kumar PL, Sharma K, Boahen S, Tefera H, Tamo M (2011a) First report of soybean witches'-broom disease caused by group 16SrII phytoplasma in soybean in Malawi and Mozambique. Plant Dis 95:492

Kumar PL, Hanna R, Alabi OJ, Soko MM, Oben TT, Vangu GHP, Naidu RA (2011b) Banana bunchy top virus in sub-Saharan Africa: investigations on virus distribution and diversity. Virus Res 159:171–182

Legg JP, Fauquet CM (2004) Cassava mosaic geminiviruses in Africa. Plant Mol Biol 56:585–599

Legg JP, Jeremiah SC, Obiero HM, Maruthi MN, Ndyetabula I, Okao-Okuja G, Bouwmeester H, Bigirimana S, Tata-Hangy W, Gashaka G, Mkamilo G, Alicai T, Kumar PL (2011) Comparing the regional epidemiology of cassava mosaic and cassava brown streak virus pandemics in Africa. Virus Res 159:161–170

Li L, Tacke E, Hofferbert HR, Lübeck J, Strahwald J, Draffehn AM, Walkemeier B, Gebhardt C (2013) Validation of candidate gene markers for marker-assisted selection of potato cultivars with improved tuber quality. Theor Appl Genet, 126:1039–1052

Liebenberg MM, Mienie CMS, Pretorius ZA (2006) The occurrence of rust resistance gene Ur-13 in common bean cultivars and lines. Euphytica 150:365–386

Lindhout P, Meijer D, Schotte T, Hutten RC, Visser RG, van Eck HJ (2011) Towards F_1 hybrid seed potato breeding. Potato Res 54(4):301–312

Mba REC, Stephenson P, Edwards K, Melzer S, Nkumbira J, Gullberg U, Apel K, Gale M, Tohme J, Fregene M (2001) Simple sequence repeat (SSR) markers survey of the cassava (*Manihot esculenta Crantz*) genome: towards an SSR-based molecular genetic map of cassava. Theor Appl Genet 102(1):21–31

Miklas PN, Kelly JD, Beebe SE, Blair MW (2006) Common bean breeding for resistance against biotic and abiotic stresses: from classical to MAS breeding. Euphytica 147:106–131

Mori K, Sakamoto Y, Mukojima N, Tamiya S, Nakao T, Ishii T, Hosaka K (2011) Development of a multiplex PCR method for simultaneous detection of diagnostic DNA markers of five disease and pest resistance genes in potato. Euphytica 180(3):347–355

Mwanga ROM, Kriegner A, Cervantes-Flores JC, Zhang DP, Moyer JW, Yencho GC (2002) Resistance to sweetpotato chlorotic stunt virus and sweetpotato feathery mottle virus is mediated by two separate recessive genes in sweetpotato. J Am Soc Hortic Sci 127:798–806

Mwanga ROM, Ghislain M, Kreuze J, Ssemakula GN, Yencho C (2011) Exploiting the use of biotechnology in sweetpotato for improved nutrition and food security: progress and future outlook. In: Conference on agro-biotechnology, biosafety and seed systems in developing countries, Kampala, Uganda, p 25

Naderpour M, Lund OS, Larsen R, Johansen E (2010) Potyviral resistance derived from cultivars of *Phaseolus vulgaris* carrying bc-3 is associated with the homozygotic presence of a mutated eIF4E allele. Mol Plant Pathol 11:255–263

Naidu RA, Robinson DJ, Kimmins FM (1998) Detection of each of the causal agents of groundnut rosette disease in plants and vector aphids by RT-PCR. J Virol Methods 76:9–18

Namukwaya B, Tripathi L, Tripathi JN, Arinaitwe G, Mukasa SB, Tushemereirwe WK (2012) Transgenic banana expressing Pflp gene confers enhanced resistance to Xanthomonas Wilt Disease. Transgenic Res 21:855–865

Narina SS, Buyyarapu R, Kottapalli KR, Sartie A, Ali MI, Robert A, Mignouna JDH, Sayre BL, Scheffler BE (2011) Generation and analysis of expressed sequence tags (ESTs) for marker development in yam (*Dioscorea alata L.*). BMC Genomics 12:100

Odeny D, Jayashree B, Ferguson M, Hoisington D, Crouch J, Gebhardt C (2007) Development, characterisation and utilisation of microsatellite markers in pigeonpea [*Cajanus cajan* (L.) Millsp.]. Plant Breed 126:130–136

Ogwok E, Odipio J, Halsey M, Gaitan-Solis E, Bua A, Taylor NJ, Fauquet CM, Alicai T (2012) Transgenic RNA interference (RNAi)-derived field resistance to cassava brown streak disease. Mol Plant Pathol 13(9):1019–1031

Okogbenin E, Porto MCM, Egesi C, Mba C, Espinosa E, Santos LG, Ospina C, Marín J, Barrera E, Gutiérrez J, Ekanayake I, Iglesias C, Fregene MA (2007) Marker-assisted introgression of resistance to cassava mosaic disease into Latin American germplasm for the genetic improvement of cassava in Africa. Crop Sci 47:1895–1904

Ortega F, Lopez-Vizcon C (2012) Application of molecular marker-assisted selection (MAS) for disease resistance in a practical potato breeding programme. Potato Res 55(1):1–13
Ravi M, Chan SWL (2010) Haploid plants produced by centromere-mediated genome elimination. Nature 464(7288):615–618
Ribaut JM, Ragot M (2007) Marker-assisted selection to improve drought adaptation in maize: the backcross approach, perspectives, limitations, and alternatives. J Exp Bot 58:351–360
Ribaut JM, de Vicente MC, Delannay X (2010) Molecular breeding in developing countries: challenges and perspectives. Curr Opin Plant Biol 13:213–218
Ribaut JM, Delannay X, McLaren G, Okono F (2012) Molecular breeding platforms in world agriculture. In: Meyers RA (ed) Encyclopedia of sustainability science and technology, vol 9. Springer, Heidelberg, pp 6692–6720
Roderick H, Tripathi L, Babirye A, Wang D, Tripathi JN, Urwin PE, Atkinson HJ (2012) Generation of transgenic plantain (Musa spp.) with resistance to plant pathogenic nematodes. Mol Plant Pathol 13:842–851
Schafleitner R, Tincopa LR, Palomino O, Rossel G, Robles RF, Alagon R, Rivera C, Quispe C, Rojas L, Pacheco JA, Solis J, Cerna D, Kim JY, Hou J, Simon R (2010) A sweetpotato gene index established by de novo assembly of pyrosequencing and Sanger sequences and mining for gene-based microsatellite markers. BMC Genomics 11(1):604
Semon M, Nielsen R, Jones MP, McCouch SR (2005) The population structure of African cultivated rice *Oryza glaberrima* (Steud): evidence for elevated levels of linkage disequilibrium caused by admixture with *O. sativa* and ecological adaptation. Genetics 169:1639–1647
Tadele Z, Assefa K (2012) Increasing food production in Africa by boosting the productivity of understudied crops. Agronomy 2:240–283
Thiémélé D, Boisnard A, Ndjiondjop M-N, Chéron S, Séré Y, Aké S, Ghesquière A, Albar L (2010) Identification of a second major resistance gene to Rice yellow mottle virus, RYMV2, in the African cultivated rice species, *O. glaberrima*. Theor Appl Genet 121:169–179
Tripathi L, Odipio J, Tripathi JN, Tusiime G (2008) A rapid technique for screening banana cultivars for resistance to Xanthomonas wilt. Eur J Plant Pathol 121:9–19
Tripathi L, Mwangi M, Abele S, Aritua V, Tushemereirwe WK, Bandyopadhyay R (2009) Xanthomonas wilt: a threat to banana production in East and Central Africa. Plant Dis 93:440–451
Tripathi L, Mwaka H, Tripathi JN, Tushemereirwe WK (2010) Expression of sweet pepper Hrap gene in banana enhances resistance to *Xanthomonas campestris* pv. *musacearum*. Mol Plant Pathol 11:721–731
Tryphone GM, Chilagane LA, Protas D, Kusolwa PM, Nchimbi-Msolla S (2012) Introgression of common bacterial blight (*Xanthomonas axonopodis* pv. *phaseoli*) resistance to common bean (*Phaseolus vulgaris* L.) adapted to Tanzania facilitated by marker assisted selection. Int J Agr Sci 2:285–290
Tumwegamire S, Rubaihayo PR, LaBonte DR, Diaz F, Kapinga R, Mwanga ROM, Grüneberg WJ (2011) Genetic diversity in white-and orange-fleshed sweetpotato farmer varieties from East Africa evaluated by simple sequence repeat markers. Crop Sci 51(3):1132–1142
Upadhyaya HD, Pundir RPS, Dwivedi SL, Gowda CLL, Reddy VG, Singh S (2009) Developing a mini core collection of sorghum for diversified utilization of germplasm. Crop Sci 49:1769–1780
Urrea CA, Miklas PN, Beaver JS, Riley RH (1996) A codominant randomly amplified polymorphic DNA (RAPD) marker useful for indirect selection of BGMV resistance in common bean. J Am Soc Hortic Sci 121:1035–1039
Varshney RK, Glaszmann J-C, Leung H, Ribaut J-M (2010) More genomic resources for less-studied crops. Trends Biotechnol 28:452–460
Varshney R, Ribaut J-M, Buckler ES, Tuberosa R, Rafalski JA, Langridge P (2012) Can genomics boost productivity of orphan crops? Nat Biotechnol 30:1172–1176

Waliyar F, Siambi M, Jones R, Reddy SV, Chibonga D, Kumar PL, Denloye S, Chinoko Y (2008) Institutionalizing aflatoxin testing in Africa. In: Leslie JF, Bandyopadhyay R, Visconti A (eds) Mycotoxins: detection methods, management, public health and agricultural trade. CABI Publishing, Wallingford, pp 359–368

Wang J, Chuang K, Ahluwalia M, Patel S, Umblas N (2005) High-throughput SNP genotyping by single-tube PCR with Tm-shift primers. Biotechniques 39:885–893

Xu K, Mackill DJ (1996) A major locus for submergence tolerance mapped on rice chromosome 9. Mol Breed 2:219–224

Chapter 7
Towards Optimizing the Impact of Tissue Culture Banana in Kenya

M.M. Njuguna and F.M. Wambugu

Abstract Banana (*Musa* spp.) is an important staple food and source of income for small-scale subsistence farmers in eastern Africa. The application of certified tissue culture (TC) planting material in smallholder farms, through a whole value-chain approach, has demonstrated significant improvement in productivity and income, confirming that the full potential of this crop is yet to be realized. Other attendant benefits to the adopters are economic, social, community, and health benefits. The smallholder TC adopters are entrepreneurs, and scaling up the TC adoption for national or regional impact will require a systems approach involving multidisciplinary teams to support them with technology development and access; technology transfer system; access to complementary inputs, credit or subsidy; infrastructural support; market linkages and conducive policy environment.

Keywords Impact • Value chain • Biotechnology • Adoption • Farmers • Scale up • Kenya

Abbreviations

AGRA	Alliance for Green Revolution in Africa
AHBFI	Africa Harvest Biotech Foundation International
EO	Entrepreneurial orientation
FFS	Farmers Field Schools
GCA	Gross cropped area
GDP	Gross Domestic Product (Kenya)
HBGMA	Highridge Banana Growers and Marketing Association
IDRC	International Development Research Centre

M.M. Njuguna (✉) • F.M. Wambugu
Africa Harvest Biotech Foundation International, Whitefield Place, 3rd Floor, School Lane, Westlands, Village Market, P.O. Box 642-00621, Nairobi, Kenya
e-mail: mnjuguna@africaharvest.org; fwambugu@africaharvest.org

F. Wambugu and D. Kamanga (eds.), *Biotechnology in Africa*, Science Policy Reports 7, DOI 10.1007/978-3-319-04001-1_7,

KARI	Kenya Agricultural Research Institute
KEPHIS	Kenya Plant Health Inspectorate Service
MTIP	Medium-Term Investment Plan (Kenya)
NARS	National Agricultural Research System (Kenya)
NBA	National Banana Association
NGO	Non-governmental organization
RF	Rockefeller Foundation
TC	Tissue culture
TCBEL	Tee Cee Banana Enterprises Limited

Background

Advances in science and agricultural technologies have delivered real benefits to farmers, processors, and consumers, and are continuing to increasingly provide new ways of tackling the challenges of reducing hunger, malnutrition, and food insecurity (CGIAR 2005). Application of biotechnology has opened new frontiers in improving food production and the incomes of smallholder farmers in developing countries. Proponents of the use of modern technology to increase production and productivity draw their inspiration from the success of the 'Green Revolution' in the latter half of the twentieth century. This resulted in a dramatic worldwide increase in food production, achieved by coupling higher-yielding plant varieties with such increasingly intensive technologies as irrigation, chemical fertilizers, and pesticides (FAO 1996).

According to Kenya's Medium-Term Investment Plan (MTIP) (Government of Kenya 2010a, b), the agricultural sector is important to the economy and has performed well in recent years, growing faster than the rural population. It directly contributes 26 % of Gross Domestic Product (GDP) and another 25 % indirectly. The sector accounts for 65 % of Kenya's total exports, as well as providing more than 70 % of informal and more than 18 % of formal employment in rural areas. The agricultural sector is therefore the mainstay of the Kenyan economy, and supports the livelihoods of the majority of the people through creating jobs and providing incomes (Government of Kenya 2010a, b).

The government recognizes that increasing food output coupled with improved access via markets is a critical means to reduce food insecurity in aggregate. Many organizations, including the Kenya Agricultural Research Institute (KARI) and Africa Harvest Biotech Foundation International (AHBFI) working in partnership with the Ministry of Agriculture, have been responding to the food production constraint by harnessing the potential of modern biotechnology through public and private partnerships and commercialization of services. The use of tissue culture (TC), an innovation used for improved banana production among smallholder farmers, has been applied for close to a decade and a half with remarkable success.

TC technology was introduced in Kenya in 1997. The studies on banana in Kenya relate to a range of subjects such as the characterization of the banana production and marketing system (Wambugu and Kiome 2001; Mbogoh et al. 2003), the processes involved in transfer of TC banana technology (Karembu and Njuguna 2000; Karembu 2007), and ex-ante assessments of its impact (Qaim 1999a, b; Qaim 2000). There are also publications on the impact, or possible impact, of transgenic or genetically modified organisms (GMOs) and gene technologies (Qaim 2001; Qaim 2003), and more recently, the social economic benefits of the TC banana by Njuguna et al. (2010).

Biotechnology innovations that tackle economically important biotic or abiotic problems hold the greatest promise for smallholder farmers in sub-Saharan Africa (De Vries and Toenniessen 2001). Studies on the role and impact of non-gene-modifying biotechnologies are few, and those that focus on the economic and social impact of non-genetically modified (GM) agricultural biotechnologies in developing countries pertain only to rice (*Oryza sativa*), maize (*Zea mays*), potato (*Solanum tuberosum*), and banana (*Musa* spp.) (FAO 2005). The literature related to banana pertains to Taiwan (Hwang and Su 1998), Nigeria (Blomme et al. 2000), Tanzania (Gallez et al. 2004), and Uganda (Lusty and Smale 2003).

The banana crop is a major source of income in Nyanza and the Central, Eastern, and Western Provinces of Kenya, and is a food item consumed in all the seven rural provinces of Kenya. Banana is used for cooking as well as for raw consumption (dessert banana). The area under production of banana in Kenya experienced a major decline in the middle of the 1990s due to severe incidence of plant diseases and nematodes. Given this situation, the introduction of tissue culture (TC) banana plantlets and associated technologies was considered as a desirable option. The introduction of new banana technology required a missionary approach, involvement of a large number of organizations, and participation of smallholder farmers. The organizations involved in these effort included technology providers, government extension providers, marketing agencies, input suppliers, and project-facilitating agencies.

Trends in Banana Area and Production

Banana occupies 7.4 % of the gross cropped area (GCA), and accounts for 55 % of total area under fruits in Kenya. The trends in area, production, and yield of banana in Kenya between 1992 and 2004 are shown in Table 7.1. The area under banana declined sharply from 1993 until 1996. This decline within a span of 3 years was mainly attributed to: (a) high incidence of pests [banana weevils (*Cosmopolites sordidus*) and nematodes] and diseases such as banana leaf spot, fusarium wilt, and cigar end rot, (b) non-availability of disease-free planting material, (c) poor agronomic and plant-husbandry practices, including no pruning, and little or no use of manures and fertilizers, leading to low yields of banana, and (d) poor post-harvest handling of bananas (Wambugu et al. 2000).

When these problems received attention from researchers, agricultural extension workers, and non-governmental development organizations, and when certified

Table 7.1 Banana: area, production, and yield in Kenya (1992–2004)

Year	Area (ha)	Production (t)	Yield (t/ha)[a]
1992	76,917	985,982	12.8
1993	79,591	817,508	10.3
1994	49,575	489,537	9.9
1995	44,434	445,733	10.0
1996	45,269	500,627	11.1
1997	75,131	1,057,586	14.1
1998	75,502	1,128,297	14.9
1999	75,286	1,097,673	14.6
2000	74,308	1,027,768	13.8
2001	77,576	1,084,312	14.0
2002	78,154	1,073,001	13.7
2003	79,598	1,019,377	12.8
2004	81,673	1,036,138	12.7

Source: Ministry of Agriculture (2005)
[a]Adopted from Njuguna et al. (2010)

disease-free planting material was made available through TC, the area under banana increased after the mid 1990s. Between 1996 and 2004, the banana area in Kenya increased at a compound growth rate of 7.6 %/year. However, the major recovery in banana area occurred during 1997. Since then, the banana area in the country has increased at a modest compound rate of 1.2 %/year only. The recovery in banana area in the later half of the 1990s was accompanied by improvements in average banana yields (MOA 2005).

Structure of Banana Farms

To understand the structure of banana-production systems in Kenya, it is important to look at the size of banana holdings and the quantity of banana produced by different groups of farmers. Qaim (1999a, b) divided the banana growers of Kenya into three groups on the basis of area under banana: small-scale (<0.2 ha), medium-scale (0.2–0.8 ha), and large-scale (>0.8 ha) farmers. According to Qaim (1999a, b), 79.6 % are small-scale farms, 18.6 % are medium-scale farms, and only 1.8 % are large-scale farms. The average area under banana is 0.12 ha on small-scale farms, 0.45 ha on medium-scale farms, and 1.98 ha on large-scale banana farms. Considering the average banana holding of the three classes and the percentage distribution of number of holdings, the average size of a banana holding in Kenya as a whole is 0.21 ha. Using this average, and in view of 81,673 ha as banana area in 2004 (MOA 2005), the total number of banana farms (or farm households with banana plants) by 2006 was estimated at 380,000.

Features of Banana, Technology Generation, Diffusion, and Transfer

The setback to the banana sector of Kenya during the middle of the 1990s occurred mainly due to the traditional method of propagation of banana, based on the use of banana suckers as planting material, which was instrumental in spread of pests and diseases (Wambugu et al. 2000). The decline in yield was reported to be as high as 90 % in some areas (Qaim 1999a). Between 1997 and 2002, the Kenyan Agricultural Research Institute (KARI), working in partnership with other development agencies with financial support from RF and the International Development Research Centre (IDRC), evaluated the feasibility and appropriateness of TC technology within the farming practices of small-scale farmers.

From 2003 to 2011, the Africa Harvest Biotech Foundation International (AHBFI), in partnership with other development agencies, implemented TC banana projects in the central and eastern provinces in Kenya with support from DuPont, Rockefeller Foundation, and the Alliance for Green Revolution in Africa (AGRA). The projects focused on increasing yields and incomes by removing bottlenecks to plantlet distribution and market access, using the whole value-chain approach. The approach focused on creating awareness of the benefits of the TC banana among farmers, producing disease-free planting material, transfer of new TC banana technology to farmers, arranging for supply of needed inputs for adoption of TC banana technology, training of farmers in agronomic and post-harvest management practices, and evolving an efficient and farmer-friendly system of disposal of marketed surplus of banana. About 35,000 households were mobilized and trained. With an average household size of 6–7 members, this represented over 200,000 beneficiaries. In respect of marketing activities, two models were simultaneously put in place, in consultation with the participating farmers. The first was a marketing company with farmers as shareholders, Tee Cee Banana Enterprises Limited (TCBEL), and the second was organizing the farmers into marketing groups, with the setup of collection centers and the establishment of tie-ups with potential buyers/traders. In addition, AHBFI established networks of farmer groups with other development partners, including the public and private sector, to produce TC banana plantlets, and used mass media to widely disperse the advantages of TC banana demonstrated in the selected project areas.

Superiority of TC Banana

The farmers who adopted TC banana technology were fully convinced of the superiority of TC banana in several ways, including: (a) the availability of large quantities of clean and superior planting material, enabling them to reclaim their old banana orchards, (b) substantial reduction in losses from pests and diseases, (c) increased productivity, (d) a shorter maturing period, and (e) uniformity of bunch sizes resulting in easy coordination of marketing (Wambugu and Kiome 2001). The estimates of costs and returns from non-TC and TC banana plantations show that the establishment and annual recurring costs of TC banana plantations are considerably higher than that for non-TC banana plantations, but that the gross income from TC banana considerably exceeds that of non-TC banana, resulting in a higher net income from TC banana (Mbogoh et al. 2003). The establishment cost of TC banana plants is paid back within a year of establishment.

Smallholder TC Farmers Are Entrepreneurs

The approach where smallholder farming is viewed as a subsistence occupation has contributed to the culture of neglect and knee-jerk reaction when it comes to interventions directed towards these important segments of the African economy. There is burgeoning evidence in literature that emphasizes entrepreneurship as the primary act underpinning innovation (Amit et al. 1993; Drucker 1985; Stevenson and Jarillo 1990). A study by Njuguna (2011) revealed that smallholder farmers adopting the TC banana in Kenya had many entrepreneurial characteristics. The study assessed the entrepreneurial orientation (EO), which refers to the processes, practices, and decision-making activities used by entrepreneurs that lead to the initiation of an entrepreneurial firm (Lumpkin and Dess 1996). Using a construct with three dimensions, proactiveness, risk taking, and innovativeness (Covin and Slevin 1986, 1989, 1991; Miller 1983) common in typical conceptualizations of EO, the study established that 75 % of the smallholder TC banana adopters were in the categories of either very high or high EO, while 20 % were moderate and the remaining 5 % were in the low EO bracket. This outcome confirmed that most of the TC banana adopters are indeed entrepreneurs who are engaged in TC banana farming. This also confirms the fact that entrepreneurs can be found in all occupations, including smallholder farming.

The TC adopters demonstrated other entrepreneurial characteristics which are documented widely in literature, including use of innovation, by adopting TC technology; entrepreneurs accept the personal financial risks that go with owning a business, but also benefit directly from the potential success of the business. The farmers had shown risk-taking behaviour, which entails marshalling required resources, including those outside their sphere of control (McFadzean et al. 2005). Constrained by resource limitations (especially finance), entrepreneurs use creativity, social networking, and bargaining to obtain favours, conduct deals, and achieve action (McGrath 1997). The farmers had secured credit and mobilized external resources to enable them to adopt TC banana. They were involved in networking activities with several partner organizations that contribute to their adoption success. The smallholders are also known to approach their farming activities as a business, by keeping records and engaging in value addition.

Scaling Up TC Banana Adoption

The adoption of TC banana can be scaled up to achieve major impact in the country in alleviating poverty, food insecurity, and malnutrition of rural communities. Based on field experience acquired in the last one and a half decades of implementing TC banana programmes, the strategy should build on seven pillars to upscale the efforts for expansion of area and productivity of TC banana as shown in Fig. 7.1. The first of the seven pillars for expansion is a system for continuous generation, optimization, and delivery (technology development and access). This task naturally should be undertaken by Kenya's National Agricultural Research System (NARS) working hand in hand with the private sector players. While the public sector institutions will be involved in technology development and optimization through laboratory and field evaluation, the private sector will be engaged in mass micropropagation. The second pillar is to strengthen the technology transfer system, through information, training, and technical assistance. This will require close interaction between the technology development agents and those involved in transfer. The third pillar should be a focus on complementary inputs such as fertilizers, basic irrigation support, plant protection chemicals, and farm equipments. This will ensure optimal performance of the TC in the farmer's fields. The fourth pillar should be access to credit or subsidy to promote uptake of the innovations. The fifth pillar should be putting in place physical and institutional infrastructure for timely distribution of the TC technology, such as the hardening nurseries. This will enhance technology diffusion. The sixth pillar should relate to evolution of a marketing system to enable the banana growers to sell their surplus produce quickly, easily, and at remunerative prices. This will serve as the lubricant that will drive the adoption of the TC. The seventh and the last pillar that will be needed to for the scaling- up objective is continuous review of the policy environment that is needed for all the other six systems/foundations to function efficiently.

The role of each component is described below;

(i) Research and Technology Development

TC Banana Technology development and access should be anchored within the NARS Programme. The programme should undertake all encompassing research on all aspects of banana production including varietal trials, introductions, evaluations, and optimization for the country's different agroclimatic conditions, pest and disease surveillance work, and perfecting agronomic practices, including harvesting and post-harvest techniques on a regular basis. The research work should also recognize TC banana as a part of a farming system that includes poultry, maize, beans, and milk (goat or cow), on a subsistence small-scale farm. Based on regular experiments, NARS should bring out a package of recommended practices for different counties which should be updated on a regular basis. This function should naturally be led by KARI, and include public universities where a team of scientists drawn from all relevant disciplines can be networked to participate in banana

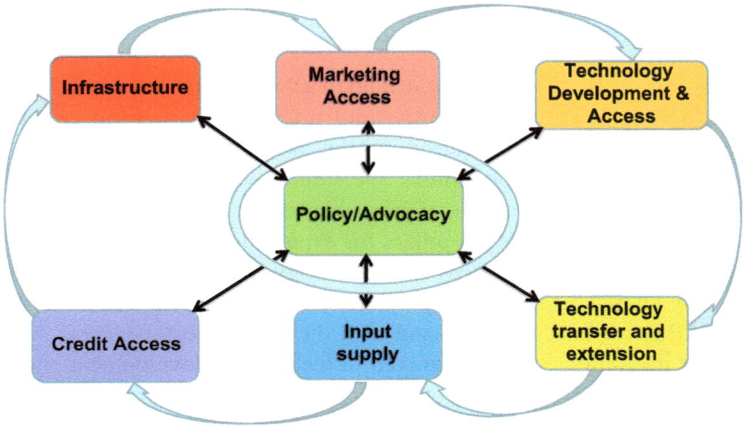

Fig. 7.1 Key components for TC banana scale up

research, technology development, and access. The programme should include representatives of private sector laboratories in the country.

(ii) Technology Transfer and Extension System

This is a very important component of the system and should incorporate information, training, and technical assistance. Currently, the County Agriculture Officers of Ministry of Agriculture are mandated to do technology transfer. This function should incorporate private sector laboratories, not-for-profit organization and other entrepreneurs. The technology transfer and extension will require (a) organization of TC banana farmers into viable groups, (b) awareness creation and repeated training of farmers, (c) laying out demonstration plots, (d) conducting field days on demonstration plots, and (e) training of some farmer leaders as trainers and other related activities.

The TC banana should be an exclusive mandate of county agricultural extension service staff of Ministry of Agriculture. For counties where banana is a major crop, a county-level TC banana coordinator be appointed, assigned specific targets, and based in an organization with considerable expertise and experience in implementation of TC banana technology projects. Agencies including NGOs involved in the promotion of TC banana should receive direct financial support from the government to scale up their activities, and farmer groups should be organized into TC Banana Growers Groups and be provided with the necessary functional support for establishing demonstration plots, hardening nurseries, input stocking depots, and revolving funds.

(iii) Input Supply System

Access to complementary inputs such as fertilizers, pesticides, and water are known to promote adoption. For successful diffusion of TC banana technology, availability of inputs and services needed for adoption of the technology is very critical. The inputs should be available in time, at reasonable prices/charges, and at convenient locations. The inputs that are critical for TC banana technology are TC banana plantlets, fertilizers, water, and plant protection chemicals. Organic manure is also an important input, but it can be arranged by the farmers themselves. Availability of TC banana plantlets requires hardening nurseries in the neighborhood, or an efficient system of transport from distant nurseries. For scale up, every division in banana-growing regions should have at least a hardening/distribution nursery to promote access. A network of agro dealers who can stock fertilizer and plant protection chemicals within the reach of farmer groups should be supported.

(iv) Credit/Subsidy system

Several studies argue that the need to undertake fixed investments may prevent small farms from adopting innovations quickly. Capital in the form of either accumulated savings or access to capital markets is required to finance many new agricultural technologies. Thus, differential access to capital is often cited as a factor in differential rates of adoption. This is the case with technologies that require a substantial initial investment. Since the adoption of the TC requires significant capital outlay, lack of credit can be a major constraint for smallholder households. Credit availability is critical for the scale up of TC banana technology. Each group can be supported to establish a revolving fund which should be managed by the group members, particularly in supporting new adopters to purchase the TC plantlet.

(v) Infrastructural support

Action to address the Kenyan agricultural crisis cannot await the achievement of ideal enabling rural infrastructure and trade-related capacities for improved market access. For agricultural development to be achieved, there is need to develop a comprehensive infrastructure to address transportation, and reduce the post-harvest losses. Development of infrastructure, particularly roads, is extremely important in increasing access to markets and reducing the cost of marketing.

(vi) Marketing System

As has been demonstrated by the success of the 'whole value chain' approach of Africa Harvest, putting in place an efficient system of marketing of TC banana is very critical for the successful up-scaling of TC banana technology. There are two important components of an efficient marketing system that reduces physical losses in marketing, increases competition, and provides quick market clearance for the TC banana growers at remunerative prices. These components are scale of marketing and technological inputs in marketing. The economies of scale in marketing can be achieved by group-marketing, rather than marketing by individual farmers. Group marketing increases the bargaining power of farmers and makes it financially feasible to introduce improved technology in marketing functions. Group marketing also reduces the price risk of banana growers.

The projects implemented by Africa Harvest have demonstrated the feasibility of a marketing company (TeeCee BEL) with farmers as shareholders. An alternative of setting up of marketing centres/collection centres where buyers and sellers are brought together has also been demonstrated by the Africa Harvest TechnoServe project; however, in the TechnoServe model there is no institutional framework. In fact, there could be other models of, say, farmers marketing cooperatives at 'location' level federated into larger cooperatives at the divisional and national level. Each centre should be provided with necessary physical and institutional infrastructure, which may include platforms, weighing balances, store space, packing house, cold store, refrigerated vans, and transport vehicles, apart from managerial services. Marketing development should work towards the manufacture of banana products to provide an alternative outlet when the supply tends to exceed demand for fresh bananas. This kind of activity can take place at farmers' group level.

(vii) Conducive Policy Environment

A favorable policy environment is the final pillar, and the linchpin of genuine transformation of the banana industry in Kenya. One of the positive developments that has a bearing on policy related to TC banana technology is that the Government of Kenya in late 2006 adopted a comprehensive 'National Biotechnology Development Policy 2006' with a view to guiding research, development, and trade in biotechnology products. In 2010, the Government adopted the biosafety bill that also provides opportunities for research in advanced technologies. The adoption of these policies has cleared the way for fast-tracking the development and application of biotechnology for the benefit of Kenya's population. In this framework, it will be prudent for the government to include TC banana as one of the priority agricultural commodities in line with maize, coffee, tea, and exportable fruits.

There are several policy issues that will need to be addressed urgently. First, there is a need for both the National and County Governments to invest in infrastructure such as road network, banana collection centers, cold rooms, storage facilities, hardening nurseries, and input supply depots. These will ensure banana-growing is done as a business and not a subsistence occupation. The credit policy should be made farmer-friendly. The Kilimo Biashara programme is a step in the right direction, as it gives a preferential interest rate to farmers. The Kenya Plant Health Inspectorate Service (KEPHIS) should put in place regulations, standards, guidelines, and an enforcement mechanism for TC plantlets produced by public or private laboratories. This will entail a clear system of inspection to ensure planting materials being produced and distributed to farmers meet acceptable standards of less than 2 % somachromal variation (mutants) and are virus-indexed. Other policy issues may include the control of diseases such as fusarium wilt, in which farmers will be required to destroy infected banana plants to minimize the spread in any particular area.

Economic and Social Impact of TC Banana

1. *Economic impact.* Banana occupies a distinct place in the national as well as in the household economy of Kenya. Nutritionally, banana stands out among other fruits because of its richness in carbohydrates, vitamins, and minerals (Wambugu and Kiome 2001). In Nyanza and the Central, Eastern and Western Provinces of Kenya, smallholder households grow and consume bananas as one of the staple foods. Nearly 83.5 % of total output of banana comes from small-scale farmers owning up to 0.5 ha of banana land (Qaim 1999a). Banana-growing families consume banana at an average rate of 300 mg/capita/day as opposed to 60 mg/capita/day by the rest of the population. As well as being a source of nutrition, banana is a reliable and regular source of cash income to around 380,000 rural families (Njuguna et al. 2010).

Further, the study done by Njuguna et al. (2010), showed that banana had significant impact at household, national, and community level.

(a) *Household level impact.* Table 7.2 shows annual income generated by 1 acre of TC over a period of 5 years. This confirms that an investment of TC in 1 acre generates a net profit of $470, which represents the free cashflow after deduction of expenses. This amount increases to $ 2,224 from the second to fifth years, when the farmer may expect some marginal reduction in yield as the orchard begins to get old. However, with good management, the high productivity can be sustained for 8–10 years. This is significant income for a household where income per capita is US $1–2. The growing of the TC banana can be a powerful tool for pulling communities out of poverty.

(b) *National level impact.* The impact of the revival of the banana sector in Kenya through the introduction of TC banana technology in tackling problems of poverty, food insecurity, and malnutrition has been immense. This is evident from the following:

- Based on figures drawn from Table 7.1, the TC banana helped recovery from the banana-industry setback that occurred during the middle of the 1990s. The area under banana, which had gone down to 46,426 ha during 1996, increased to 79,808 ha during 2004, and is likely to be around 82,000 ha during 2006. It is not just that all this additional area came under TC banana, but that the entire campaign of TC, through the key projects and initiatives, provided a ray of hope to the banana growers. An increase in the banana area of around 35,574 ha within 10 years equates to an additional net income of Ksh 5,648 million ($ 1 = Ksh 70) accruing to 380,000 banana growers.
- The additional income or increased access to one of the staple foods to small-scale-farm families has helped in improving the food security, nutritional levels, and economic status of the rural poor in Kenya.
- Banana production has provided a cash-income security to the poor farmers because it provides an almost continuous income flow throughout the year, even under low-input regimes. In addition, banana suckers and leaves are

Table 7.2 Income to household from 1 acre of TC banana

Particulars	First year	Subsequent years up to fifth year
Establishment cost (US$)	1,103	–
Annual operational cost (US$)	532	532
Total cost (US$)	1,635	532
Yield per acre (tonnes)	11.48	15.03
Price per tonne (US$)	183	183
Gross income (US$)	2,105	2,756
Net income to the farmer (US$)	470	2,224

used as animal feed, especially during dry seasons when no other source of fodder is available.

- Apart from the additional income that accrued to TC-banana-growing families, introduction of TC banana and consequent revival of the banana economy had a multiplier economic impact on rest of Kenya's economy by providing employment and business opportunities to village assemblers, wholesalers, urban retailers, transporters, laborers in wholesale markets, manufacturers of packaging materials, and agricultural labor households.
- Additional production of banana also impacted banana consumers by way of lower real prices of banana, which otherwise would have prevailed at levels higher than the existing current level of banana retail prices.

In brief, the overall economic impact of TC banana resulted in benefits for all sections of Kenya's population, and the economic benefits derived by small-scale, resource-poor farmers has been substantially more than the benefits accruing to other sectors. Our study further established that the major drawback in the use of tissue culture banana was increase in the initial orchard establishment costs, water, and household labor requirements.

2. *Social impact*. The social impact of the TC-banana project has reflected upon the adopter families, at the level of households and communities, as well as in terms of gender relations. At the household level, the impact is evident in the following ways:

    - A TC-banana plantation is considered by farm families as an important additional economic asset and security for the family.
    - An increase in banana production at the farm level has increased food security at the household level. Farm families who have adopted TC banana did not require food aid for the first time in their lives when there was a drought and when food aid was required in the area.
    - Malnutrition among members of the banana-growing households reduced, owing to the additional income that was used for the purchase of other foods, leading to diversity in diets.
    - Adoption of TC banana has led to the economic empowerment of women because in the majority of households, banana produce and income belongs to the domain of women.

- A higher income has allowed the family to improve other quality of life indicators by way of payment of secondary school fees for the children, improved housing, and diversification of income through taking up of other supplementary enterprises such as poultry.
- Decision making and dynamics within the family changed considerably with the increased income in several families.
- While monetization of banana could have the effect of lowering household consumption, the practice of selling TC banana by grade always left some banana that was not purchased by buyers. This lower-grade banana was used for home consumption. Thus, there is no evidence of decreased household consumption. In fact, increased banana production has led to increased consumption at the household level in almost all banana-growing households.

At the community level, the impact is visible in the following form:

- The formation of cohesive farmer groups has empowered the groups to address not only agronomic issues related to banana but also issues of other community interests. The farmer groups have been effective in addressing anti-social behavior within the community.
- Banana-grower group activities have provided an entry point to the development agencies for other development activities at the community level.
- The requirement of the group to support members for credit access has increased the accountability of the members to the community, thus further increasing cohesiveness among the families.
- The collective voice for community improvement has been always important, and the formation of groups has further influenced the management of community development funds.

The TC-banana adopter families revealed a distinct empowerment of farm women. In the majority of cases, farm women played a more proactive role in adoption of the new TC technology. Membership of men to women in TC-banana project groups was approximately in the ratio of 1:1. Empowerment of women through TC banana is revealed from the following:

- Improved banana production has contributed to household welfare, especially for women and children, because average access and control of income from banana sales showed higher control by women. Women contributed about 33 % of labor requirements for banana production, but the control of women over banana income was higher.
- Projected additional income to the family after adoption of TC banana is reflected in an increase in disposable income for the family, over which women have more discretion to spend. A substantial proportion of income from banana sales goes to purchase other items of food by the women.
- Banana has improved the nutrition value of household diets, and hence improved general health and productivity of the households, including women and girls.
- Banana sales have been used to improve children's education, as many households have paid school fees from the sale of bananas.

- Many families have been able to construct good houses from the sales of bananas, benefiting housewives.
- Many families, from banana sales, have acquired assets such as mobile phones, bicycles and consumer durables for the family.

Effectiveness of Whole Value-Chain Approach

The whole value-chain approach employed by Africa Harvest has been successful, and can be replicated in other countries in the region with similar success. Value chain has the advantage of being an integrated approach and quite effective in several ways, some of which include;

- *Farmer engagement:* the value chain has been very effective in awareness creation among the potential adopters. Information packaged in flyers and brochures in simplified form has empowered TC banana adopters to make informed decisions on production, marketing, and processing. The approach is cost-effective in training farmers through Farmers Field Schools (FFS) on all aspects of TC banana production technology from land preparation, digging holes, weeding, and de-suckering to harvesting and post-harvest handling techniques. It provides farmers with a platform for participatory project planning, while facilitating their involvement in the regular monitoring and evaluation of the project.
- *Farmer organization development:* the approach has been effective in group formation and management. Once formed, the groups are trained to form committees that that deal with training, marketing, production, and processing and dispute resolution. They are supported to develop constitutions, call for regular meetings, and are trained on group dynamics. This enhances cohesion and management of the groups. These groups also become entry points for other innovations and development activities.
- *Accessing innovation:* the value-chain approach has been quite successful in helping farmers access TC banana planting materials. The planting materials, sourced from private laboratories that mainly located around Nairobi, are distributed through regional hardening nurseries, some located over 400 km away, thereby facilitating access.
- *Enhancing the affordability of innovation:* the value-chain approach has had a significant contribution in making TC banana plantlets affordable. Some of the projects funded by development partners have a subsidy programme inbuilt in the project design, to offer farmers a discount ranging from 20–50 %. Further, setting-up of regional hardening nurseries in areas far from the TC laboratories has reduced transportation costs, making the plantlets more affordable. Finally, some TC banana programmes have entered into partnership with micro-finance institutions that have provided loans to TC adopters with flexible repayment schedules.

- *Farmer capacity-building:* the value-chain approach has supported comprehensive training in good agronomic practices and post-harvest management that has reduced loses significantly.
- *Enhanced access to the market:* the value chain has addressed marketing issues through facilitating farmer associations, e.g., Highridge Banana Growers and Marketing Association (HBGMA) and the National Banana Association (NBA), all of which have a special focus on marketing. The farmer-owned company (Tee Cee BEL) has also facilitated increased access to the market. Farmers have been able to be rewarded through premium prices paid for their produce.

The whole value-chain approach employed by Africa Harvest has been effective in addressing most of the bottlenecks normally associated with the diffusion of agricultural technologies. The model can be replicated in other agricultural value chains with considerable success.

Conclusions

TC-banana technology initiatives in Kenya have demonstrated that TC technology is appropriate and manageable by small-scale farmers, and that it has led to an increase in the quality and productivity of banana. The increase in the yield of banana not only satisfies home consumption, but also creates surpluses for sale in the prime marketplace. Income from this activity can reduce poverty and upgrade the social welfare of rural communities and families. There is evidence that farmers have come out of the poverty trap as a direct result of the TC-banana project. The whole value-chain process that has resulted in participating farmers moving out of the poverty trap consists of awareness creation, farmers' group organization, technology transfer and training, establishment of a distribution system of plantlets and inputs, establishment of a financing system, and provision of market linkages. The whole value-chain model links entrepreneurs and companies operating throughout the value chain, from the provision of seeds (TC plantlets) to the production in farmers' fields and all the way to the final marketing when consumers purchase a finished product. The TC-banana project whole value-chain approach can also be adapted and applied to other vegetatively propagated crops, such as pineapple (*Ananas comosus*), sweet potato (*Ipomoea batatas*) and cassava (*Manihot esculenta*), because they face similar propagation and distribution challenges.

References

Amit R, Glosten L, Muller E (1993) Challenges to theory development in entrepreneurship research. J Manage Stud 30(5):815–834

Blomme G, Dochez C, De Schutter B, Gockowsk J, Gold C, Green K, Hartman J, Hauser S, Hughes J d'A, Nolte C, Okech S, Pillay M, Ragama P, Speijer P, Tenkouano A, Tshiunza M,

Ude G, Vuylsteke D (2000) Project 7: improving plantain- and banana-based systems. In: International Institute of Tropical Agriculture (IITA) Annual Report 1999. IITA, Ibadan, pp 59–61
Consultative Group on International Agricultural Research (2005) Science for agricultural development: changing contexts, new opportunities. Science Council Secretariat, Rome
Covin JG, Slevin DP (1986) The development and testing of an organizational-level entrepreneurship scale. In: Ronstadt R, Hornaday JA, Peterson R, Vesper KH (eds) Frontiers of entrepreneurship research. Babson College, Wellesley, pp 628–639
Covin JG, Slevin DP (1989) Strategic management of small firms in hostile and benign environments. Strateg Manage J 10(1):75–87
Covin JG, Slevin DP (1991) A conceptual model of entrepreneurship as firm behavior. Entrep Theory Pract 16(1):7–25
De Vries J, Toenniessen G (2001) Securing the harvest: biotechnology, breeding and seed systems for African crops. CAB International, Wallingford
Drucker PF (1985) Innovation and entrepreneurship: practice and principles. Heinemann, London
Food and Agriculture Organization of the United Nations (1996) Rome declaration of world food security, from the world food summit. Food and Agriculture Organization of the United Nations, Rome
Food and Agriculture Organization (FAO) (2005) Annotated bibliography on the economic and socio-economic impact of agricultural biotechnology in developing countries. FAO, Rome
Gallez A, Runyoro G, Mbehoma CB, Van den Houwe I, Swennen R (2004) Rapid mass propagation and diffusion of new banana varieties among small-scale farmers in North Western Tanzania. Afr Crop Sci J 12:7–17
Government of Kenya (2010a) Agricultural sector development strategy. Government Press, Nairobi
Government of Kenya (2010b) Growth and Food security through increased agricultural productivity and trade- a medium-term investment plan for Kenya's agricultural sector. Government Press, Nairobi
Hwang SC, Su HJ (1998) Production of virus-free banana plantlets in Taiwan. Taiwan Banana Research Institute, Chiuju, Pingtung
Karembu M (2007) Enhancing the diffusion of tissue culture banana to small-scale farmers in Kenya. Tissue culture banana policy brief. ISAAA, Ithaca
Karembu M, Njuguna M (2000) Institutional issues in biotechnology applications: concepts and empirical evidence from Kenya. In: Qaim M, Krattiger AF, Von Braun J (eds) Agricultural biotechnology in developing countries: towards optimizing the benefits for the poor. Kluwer Academic, Boston, pp 175–188
Lumpkin GT, Dess GG (1996) Clarifying the entrepreneurial orientation construct and linking it to performance. Acad Manage Rev 21(1):135–172
Lusty C, Smale M (2003) Assessing the social and economic impact of improved banana varieties in East Africa. Proc. interdisciplinary research design workshop. Kampala, Uganda 7–11, November. IFPRI, Washington
Mbogoh S, Wambugu F, Wakhusama S (2003) Socio-economic impact of biotechnology applications: some lessons from the pilot tissue-culture (TC) banana production project in Kenya, 1997–2002. Proceedings of the 25th International Conference of Agricultural Economists (IAAE), Durban, 16–22 Aug 2003, pp 1084–1094
McFadzean E, O'Loughlin A, Shaw E (2005) Corporate entrepreneurship and innovation part 1: the missing link. Eur J Innov Manage 8(3):350–372
McGrath RG (1997) The parsimonious path to profit. In: Birley S, Muzyka D (eds) Mastering enterprise. Financial Times/Pitman, London
Miller D (1983) The correlates of entrepreneurship in three types of firms. Manage Sci 29(7):770–791
Ministry of Agriculture (2005) Strategic plan 2005–2009, pp 1–81
Njuguna M, Wambugu F, Acharya S, Mackey M (2010) Socio-economic impact of tissue culture banana (Musa Spp.) in Kenya through the whole value-chain approach. In: Dubois T,

Hauser C, Staver D, Coyne D (eds) Proceedings of the international conference on banana and plantain in Africa: harnessing international partnerships to increase research impact, 5–9 Oct 2008. Acta Horticulturae, Mombasa, pp 77–86

Njuguna MM (2011) Assessment of the role of social capital in the adoption of agricultural innovations among smallholder farmers: a case of tissue culture banana in Kenya. Ph.D. thesis. Kenyatta University, Kenya

Qaim M (1999a) Assessing the impact of banana biotechnology in Kenya. ISAAA Brief No. 10. ISAAA, Ithaca

Qaim M (1999b) A socioeconomic outlook on tissue culture technology in Kenyan banana production. Biotechnol Dev Monit 4:18–22

Qaim M (2000) Biotechnology for small-scale farmers: a Kenyan case study. Int J Biotechnol 2:174–188

Qaim M (2001) A prospective evaluation of biotechnology in semi-subsistence agriculture. Agric Econ 25:165–175

Qaim M (2003) Bt cotton in India: field trial results and economic projections. World Dev 31:2115–2127

Stevenson HH, Jarillo JC (1990) A paradigm of entrepreneurship: entrepreneurial management. Strateg Manage J 11:17–27

Wambugu F, Karembu M, Njuguna M, Wanyangu SW (2000) Biotechnology to benefit small-scale banana producers in Kenya. Global Development Network, New Delhi

Wambugu F, Kiome R (2001) The benefits of biotechnology for small-scale banana farmers in Kenya. ISAAA Brief No. 22. ISAAA, Ithaca

Part III
Technologies in Development

Chapter 8
The Use of African Indigenous Genes in the Development of Transgenic Maize Tolerant to Drought and Resistant to Maize Streak Virus

Jennifer A. Thomson, Sagadevan G. Mundree, Dionne M. Shepherd, and Edward P. Rybicki

Abstract When developing a plant with resistance to biotic and abiotic stresses and utilizing genetic engineering, why should scientists limit themselves to genes from known organisms? Why not test those from indigenous species that might have unique properties? In this chapter we describe the use of indigenous genes for the development of crops important to Africa. The first is maize tolerant to drought, a situation which appears to be worsening on the continent, and the second is maize resistant to the African-endemic maize streak virus. The genes for drought tolerance were derived from the resurrection plant, *Xerophyta viscosa*, which survives even when it contains only 5 % of its relative water content. The plant can be 'resurrected' within 80 h of receiving moisture. Two methods were used to identify potential genes of interest. The first was complementation by functional sufficiency in *Escherichia coli*, resulting in the isolation of *XvSap1* (which was found to code for a membrane-associated signalling protein) and *XvAld*, coding for aldose reducatase which converts glucose to sorbitol, an osmoprotectant. The second method was differential screening of expression libraries resulting in the isolation of *XvPrx2*, which codes for an antioxidant peroxiredoxin, and *XvG6*, which codes for a stress-responsive regulatory protein. Other genes isolated, tested, and not used further are also mentioned. For resistance to maize streak virus, the approach of pathogen derived resistance was used, resulting in the isolation of dominant negative mutants of the viral replication associated protein gene, *rep*. In a refinement of this approach, a virus-inducible version of the mutants was developed as well as an siRNA approach. As the development of transgenic maize is a lengthy process, the genes were first tested in model systems. For drought tolerance the

J.A. Thomson (✉) • D.M. Shepherd • E.P. Rybicki
Department of Molecular and Cell Biology, University of Cape Town, Private Bag, Rondebosch, South Africa
e-mail: Jennifer.thomson@uct.ac.za

S.G. Mundree
Centre for Tropical Crops and Biocommodities, Queensland University of Technology, Brisbane, Australia

F. Wambugu and D. Kamanga (eds.), *Biotechnology in Africa*, Science Policy Reports 7, DOI 10.1007/978-3-319-04001-1_8,

model plants were *Arabidopsis* and tobacco while for virus resistance black Mexican sweetcorn in tissue culture and transgenic *Digitaria sanguinalis*, an MSV sensitive grass, were used. Cassettes of the genes shown to be effective, including inducible systems for both drought and virus resistance, were introduced into maize and results are presented. The paper concludes with a discussion on how to bring these products to the farmers' market, in Africa.

When scientists decide to develop a crop with a specific trait, using genetic engineering, they can choose the relevant genes from a variety of different sources. The question arises, does one stick to tried and tested genes from known organisms, or does one consider those from indigenous species that might be able to solve the problem more efficiently? In this chapter, we look at two examples of how that latter approach can be used. The first is drought tolerance in maize where the resurrection plant, *X. viscosa*, was chosen for its ability to withstand the loss of 95 % of its water content and readily 'resurrect' upon watering. The second is an already proven approach, namely that of using a pathogen to derive resistance to it, and the example is the African indigenous maize streak virus.

Keywords Drought tolerance • Transgenic • Virus resistant • Maize • Biotic • Abiotic

Abbreviations

AATF	African Agricultural Technology Foundation
BMS	Black Mexican sweetcorn
cDNA	Complementary DNA
CYMMIT	International Maize and Wheat Improvement Centre
MSD	Maize streak disease
MSV	Maize streak virus
NADPH	Nicotinamide adenine dinucleotide phosphate
OP	Open pollinated
PRP	Proline-rich protein
Prx	Peroxiredoxin
PTGS	Post-transcriptional gene silencing
RT-PCR	Real-time polymerase chain reaction
RWC	Relative water content
UCT	University of Cape Town
WEMA	Water-efficient maize for Africa

Drought-Tolerant Maize

Drought is nothing new in Africa, but the situation appears to be worsening. Since mid-2011 a severe drought, said to be the worst in 60 years, has been affecting the entire East Africa region. This is causing major food crises in several countries with refugees fleeing from affected areas. One of the crops affected is maize, the most widely grown staple crop in Africa, with more than 300 million people depending on it for their main food source. Although conventional plant breeding has had some success in the development of drought tolerant maize, these processes rely on genetic information from the same or closely related species. In contrast, genetic engineering can access genes from any organism that could confer such tolerance.

Using a gene, *cspB*, which codes for a cold shock protein, from the bacterium, *Bacillus subtilis*, Monsanto has developed potentially useful drought tolerant maize varieties (Castiglioni et al. 2008). They have donated the gene, royalty free, to the Water Efficient Maize for Africa (WEMA) project which is managed by the Nairobi-based African Agricultural Technology Foundation (AATF). The International Maize and Wheat Improvement Centre (CYMMIT), which has a branch in Kenya, has introduced the gene into African maize varieties, and these are being tested by the consortium partners in South Africa, Kenya and Uganda.

The drought-tolerant research group at the University of Cape Town (UCT), however, decided to look to a different organism as the source of genes to develop this trait in maize. Although the general response to abiotic stress is similar in all plants, there is a group known as 'resurrection plants' that have developed mechanisms which enable them to withstand severe water deficit. These plants are unique in their ability to tolerate the drying of their vegetative tissues. Resurrection plants can lose over 90 % of their water content, survive in their dried state for prolonged periods and then resume active life when water becomes available again (Sherwin and Farrant 1996). It is thought that two basic mechanisms exist which allow desiccation-tolerant plants to survive such deprivation. The first involves the protection of cellular integrity through inducible and constitutive mechanisms, while the second involves the repair of desiccation or rehydration-induced damage. However, both mechanisms are probably employed for desiccation tolerance with different plants utilizing one strategy more than the other (Oliver and Bewley 1997).

The desiccation-tolerant resurrection plant *Xerophyta viscosa baker* (Family Velloziaceae) can be dehydrated to 5 % relative water content (RWC), and upon rewatering the desiccated plant rehydrates completely within 80 h, resuming full physiological activities (Sherwin and Farrant 1996; Fig. 8.1). Like most resurrection plants, *X. viscosa* grows in shallow soils on rocky outcrops at high altitudes where there is little shade. In addition to water deficit stress, night temperatures are close to 4 °C. Upon drying, it dissembles its thylakoid membranes, loses its chlorophyll (poikilochlorophyllous), and the leaf blades fold in half along the midrib, with only the abaxial surface being exposed to light. The leaves initially turn yellow and then dark purple due to the accumulation of anthocyanins, when in a more advanced dry state. The abaxial surfaces have a reflective sticky coating, which may serve to reduce light absorbed by the leaf. The activities of the three common antioxidant enzymes, ascorbate peroxidase, glutathione reductase and superoxide dismutase, increase during dehydration (Sherwin and Farrant 1998). We reasoned that a plant having such interesting metabolic

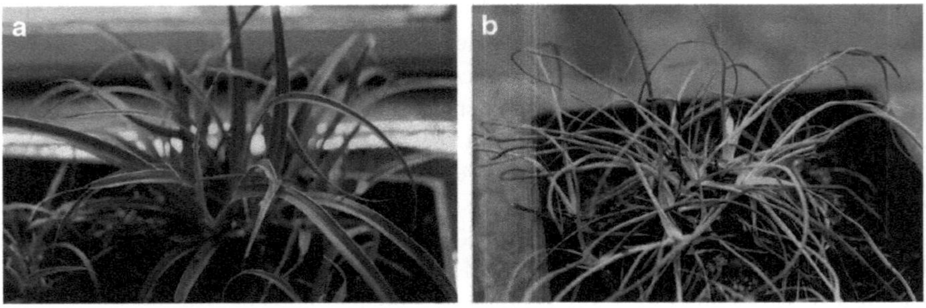

Fig. 8.1 *X. viscosa* plants (**a**) hydrated (**b**) dehydrated

Fig. 8.2 *X. viscosa* plants growing on a hillside in the Drakensberg Mountains in KwaZulu–Natal, South Africa

adjustments to dehydration should provide us with numerous potential genes for our project.

A search for plant specimens took the research group into the Drakensberg Mountains in KwaZulu-Natal, South Africa. There we found numerous specimens growing on rock crevices and faces, where they were obviously thriving despite the shallow soils in which they were growing (Fig. 8.2). We brought some of these back to our glasshouse at the University of Cape Town for cultivation, and after numerous attempts were finally able to set seed and grow new plants, obviating the need to return annually to the Drakensberg for fresh plants.

Maize Streak Virus Resistance

The second example in this indigenous gene approach is that of maize streak virus (MSV), the causal agent of maize streak disease (MSD), which is one of the major biological threats to food security in sub-Saharan Africa, where the disease is

Fig. 8.3 MSV symptoms on a maize plant (**a**) and (**b**) in a field of maize

endemic (Fig. 8.3). It is transmitted by the leafhopper, *Cicadulina mbila* Naudé. MSV is a geminivirus which, unlike most plant viruses so far discovered, has DNA and not RNA as its genetic material. There is evidence that coat protein-mediated resistance, effectively used for the latter group, would not be successful for viruses such as MSV which replicate in the nucleus. Therefore, the MSV group at UCT decided to use a variety of different approaches to solve this problem (see section "Testing of Genes in Model Systems for Drought Tolerance").

Identification of Potential Genes for Drought Tolerance: Screening *X. viscosa* Gene Libraries

Screening Using Complementation by Functional Sufficiency in Escherichia coli *and Genes Isolated*

Like most bacteria, *E. coli* can tolerate and grow in media whose osmolarity corresponds to about 0.7 M NaCl (Gowrishankar 1985). In the face of decreased turgor, the bacteria are known to undergo changes in the cell envelope structure, facilitating the accumulation of compatible solutes so that turgor is restored (McLaggan et al. 1994). However, in the absence of a specific transport system, as in the *E. coli* (*srl*::Tn10) mutant strain, any increase in osmolarity of the growth medium (such as 1.25 M sorbitol) would result in death of the cells, as this osmoticum places an osmotic stress on the cells. When genes whose products can allow this osmotic sensitivity to be overcome are introduced into these cells, the process is called "complementation by functional sufficiency" (Mundree 1996).

A cDNA library was constructed using RNA extracted from *X. viscosa* leaves that were at 85 %, 37 %, and 5 % RWC. The RNA was pooled and used as a template for cDNAsynthesis. The phagemid library was used to infect *E. coli* (*srl*::

Tn10) cells and grown on 1.25 M sorbitol. Among the genes isolated by this technique were *XvSap1* and *XvAld*.

XvSap1 Codes for a Membrane-Associated Signalling Protein

One of the clones isolated was designated *XvSap1*, standing for stress-associated protein. Nucleotide sequencing showed that the gene coded for a basic protein of 264 amino acids with a molecular weight of 29.6 kDa (Garwe et al. 2003). A motif search revealed that the protein has two prokaryotic membrane lipoprotein lipid attachment sites. A hydropathic plot predicted a protein rich in hydrophobic residues with at least six transmembrane helices and two prokaryotic membrane lipid attachment sites, suggesting that *XvSap1* is likely to be an integral membrane protein.

A computer search of protein sequence databanks revealed that the protein showed 49 % identity to WCOR413, a cold-responsive protein isolated from wheat, and between 25 % and 56 % identity to cold associated proteins identified in *Arabidopsis thaliana*. It also has 53 % identity to a cold-associated protein from rice. Additionally, it has a region which bears 12 % identity with a K^+ potassium transporter family that is conserved across phyla.

X. viscosa plants were subjected to a variety of stresses including dehydration, heat (42 °C), cold (4 °C), high light intensity (1,500 μmol m^{-2} s^{-1} for 4 days in a phytotoron at 25 °C and 50–70 % humidity), and high salinity (100 mMNaCl), and RT-PCR was used to compare the relative transcript levels. *XvSap1*was induced by dehydration, with the transcript only appearing at 51 % and 44 % relative water content (RWC). It was not detected during rehydration of the plants. Heat, cold, high light and high salinity resulted in significant induction of the gene. The transcripts took 3 days to appear after heat shock but only 24 h after cold treatment, under which conditions they remained fairly steady for the duration of the experiment. Transcripts were induced within 48 h under high light intensity and within 24 h under salt shock (Garwe et al. 2003).

An examination of the sequence of the *XvSap1* protein revealed very few clues as to its possible functions in conferring stress tolerance. It is possible that the protein may be involved in the transport of substances or ions across the plasma membrane due to the 12 % identity with the K^+ potassium transporter family (residues 36–119; Garwe et al. 2003).

As gene expression was only induced at 51 % and 44 % RWC, it would appear that *XvSap1* is not required during the initial stages of dehydration, but is only expressed when dehydration becomes severe and the plant has dried down considerably. As the protein is likely to be an integral membrane protein, one of the roles it could play is the stabilization of membranes during the drying process (Garwe et al. 2003). Further studies by Iyer et al. (2007) confirmed this hypothesis, and indicated that the protein might also be involved in maintaining ion homeostasis as well as being associated with signal transduction under osmotic stress.

XvAld Codes for Aldose Reductase

It has been assumed that molecules and compounds synthesized and accumulated during desiccation play an important role in protection from stress. Such compounds are thought to protect intracellular components such as enzymes, membranes and other macromolecules against damage due to desiccation (Oliver and Bewley 1997). Carbohydrates are among the osmoprotective organic solutes which accumulate in higher plants. One of the genes isolated, *XvAld*, codes for an aldose reductase, part of a group of the aldo–ketoreductase superfamily. These are cytosolic, monomeric oxidoreductases which catalyze the NADPH-dependent reduction of carbonyl metabolites (Bohren et al. 1989).

When the nucleotide sequence of one of the clones isolated during complementation by functional sufficiency was determined, it was found to code for a protein with significant similarity to several aldose reductases (Mundree et al. 2002). These included *Hordeum vulgare* (66 % similarity), *Bromus inermis* (65 %), and *Avena fatua* (65 %). The protein contained one motif, IPKS, which is highly conserved among aldose reductases.

Northern blot analysis of poly(A) $^+$ RNA isolated from hydrated (100 % RWC) and dehydrated (85 %, 37 %, and 5 % RWC) *X. viscosa* leaves exhibited a single transcript of the expected size (ca. 1.2 kb). It was not present in fully hydrated leaves (ibid.).

Western blot analysis using antibodies raised against a barley aldose reductase cDNA clone by Bartels et al. (1991) showed that a soluble protein of the expected size (approximately 36 kDa molecular mass) was present in dehydrated *X. viscosa* leaves (ibid.).

Enzyme assays on protein extracts from *X. viscosa* showed that in the former there was a sixfold increase in aldose reductase activity as the RWC decreased from 100 % to 15 %. As a control, *Sporobolus stafianus*, a non-resurrection grass, showed relatively little change in enzyme activity with changing RWC.

Differential Screening of Expression Libraries and Genes Isolated

RNA was extracted from *X. viscosa* leaves that were at 85 %, 37 %, or 5 % RWC. The RNA was pooled and used as a template to construct a cDNA library in *E. coli*. Plasmid DNA from isolated colonies was slot-blotted in duplicate onto nylon membranes. RNA was isolated from hydrated (100 % RWC) and dehydrated (37 %) RWC leaves and reverse-transcribed into DNA incorporating [^{32}P] dCTP. This was used to probe the membranes separately, and differentially expressed genes were identified following autoradiography (Ndima et al. 2001). Genes were also isolated from *X. viscosa* leaves that had been subjected to 4 °C for 60 h. Genes from both libraries were subjected to DNA sequence analysis, and the Genbank database was searched for sequence similarities. Genes coding for proteins that could be involved in desiccation stress tolerance were studied further.

XvPrx2 Codes for an Antioxidant Peroxiredoxin

A stress-inducible gene designated *XvPrx2*, which has 77 % identity with the *Oryza sativa* orthologue, was chosen for further analysis due to its apparent role in oxidative stress as a type II peroxiredoxin (Prx). Prxs reduce hydrogen peroxides and alkyl peroxides to water and alcohols, respectively, by using reducing equivalents derived from thiol-containing donor molecules. The *1-CysPrx* transcript of *X. viscosa* accumulates in tissues under abiotic stresses such as dehydration, heat, high light intensity, and salinity. This indicates that the protein has a significant role in drought tolerance (Dietz 2011).

Northern blot analyses were performed to determine whether *XvPrx2* is stress-inducible. The transcript was detected after low temperature (4 °C), high temperature (42 °C), dehydration, salt (150 mMNaCl), and high light intensity (1,500 μmol m^{-2} s^{-1}) treatments. Most of these changes were mirrored by some levels of protein increase, although these fluctuated between treatments and over time. It is possible that there is some post-transcriptional regulation of *XvPrx2*, which may account for the differences in the correlation of mRNA and protein levels (Govender 2006).

XvG6 Codes for a Stress-Responsive Regulatory Protein

Sequence analysis of *XvG6* showed that it had no identity to known plant genes. *In silico* analysis of the encoded protein predicted that it was a proline-rich protein (PRP), as it possessed many of the features common to such proteins. These include a signal peptide, proline repeats, a cysteine residue, and possible phosphorylation sites (Felix 2007). It also possesses a tyrosine residue present in some of the repeats, which is believed to play a role in protecting the plant against environmental stresses (Bradley et al. 1992). Western blot analysis showed that the protein levels increased during dehydration (Felix 2007).

Genes Isolated but Not Used

Some of the genes discussed here were shown not to protect transgenic plants from abiotic stresses, but some of them have not been used simply due to a lack of sufficient human capacity and funds to undertake the research.

XvPer1 Codes for an Antioxidant Peroxiredoxin (Prx)

The *XvPer1* cDNA was isolated by the differential screening method. DNA sequencing showed that it coded for a protein showing considerable similarity to other plant 1-Cys Prx homologues (Mowla et al. 2002). This antioxidant is unusual in that it is found in *X. viscosa* vegetative tissue, while most are seed-specific. It has a nuclear localization signal and has been localized to the nucleus in leaf tissue (ibid.)

The transcript was induced by abiotic stresses such as dehydration, heat (42 °C), and high light intensity (1,500 μmol photons m^{-2} s^{-1}), and when treated with abscisic acid (100 μM) and sodium chloride (100 mMNaCl). Western blot analysis correlated with the patterns of expression of the *XvPer1* transcripts. Immunofluorescence analyses revealed that *XvPer1* is localized in the nucleus of dehydrated *X. viscosa* leaf cells. These results suggested that the protein may function to protect nucleic acids within the nucleus against oxidative injury (ibid.)

Transgenic maize plants with constitutively over-expressed *XvPer1* were also produced. Unfortunately, these showed no improvement in growth under dehydration stress, although they did show an advantage over control plants when subjected to high light intensity stress (Mowla 2005). The effect was, however, not sufficient for us to pursue this gene.

XvATP1 Codes for a Vacuolar H^+-ATPase

The plant vacuole primarily maintains cellular turgor pressure along with other functions such as giving the cell shape and rigidity, increasing the cellular surface area to facilitate efficient photosynthesis, and absorption of nutrients and storage or various compounds that could be toxic to the cell if released into the cytoplasm (Taiz 1992). It also plays a vital role in maintaining ion homeostasis between itself and the cytoplasm by facilitating the functioning, among others, of vacuolar H^+-adenosine triphosphatases (Serrano and Rodriquez-Navarro 2001).

The *XvATP1* gene was isolated using the strategy of complementation by functional sufficiency in *E. coli*. The protein showed significant homology to the proteolipidc subunit of the vacuolar H^+-ATPase complex involved in supplying tonoplastic energy to plants. When water deficit is imposed on a plant, water is passively extracted from cells, concentrating solutes within, which could lead to hyperosmotic tension across the tonoplast, the vacuolar membrane. It was proposed that overexpression of this gene could help to restore osmotic equilibrium in the vacuole, which can comprise up to 90 % of the cell volume (Marais et al. 2005).

Northern blot hybridization was conducted on RNA isolated from *X. viscosa* plants subjected to dehydration, salinity and cold (−20 °C for 120 min) shock. An increase in the *XvATP1* gene transcript was seen in response to all these conditions (ibid.).

XvGolS Codes for a Galactinol Synthase

As mentioned in section "Maize Streak Virus Resistance", osmoprotectants could help to protect plants from damage suffered during dehydration. One of the genes isolated during differential screening of gene libraries was *XvGolS* which codes for a galactinol synthase. GolS enzymes represent the first step in the synthesis of the raffinose family oligosaccharides, major soluble carbohydrates occurring in the seeds and other vegetative tissues of plants (Peterbauer et al. 2002). Galactinol, an α-galactoside of myo-inositol, is an unusual molecule found exclusively in plants. Its biosynthesis is catalyzed by the galactosyltransferase, galactinol synthase, which uses myo-inositol and UDP-galactose as substrates (Sprenger and Keller 2000).

XvGolS was found to be up-regulated in the leaves of *X. viscosa* during dehydration stress (Mundree et al. 2002; Peters et al. 2007).

XvT8 Codes for a Dehydrin

Another gene isolated during differential screening, *XvT8*, was found to code for a dehydrin, as its protein exhibited 45 % and 43 % identity to similar proteins from *A. thaliana* and *Pisum sativum* respectively. Dehydrins are known to accumulate in response to a variety of stresses, and indeed transcripts of *XvT8* accumulated in *X. viscosa* plants that were exposed to heat, cold, and dehydration stresses (Ndima et al. 2001).

Identification of Potential Genes for MSV Resistance

*Dominant Negative Mutants of the Replication-Associated Protein Gene (*rep*)*

MSV has a small (2.7 kb), single-stranded circular DNA genome encoding the coat protein, movement protein, and two replication-associated proteins, Rep and RepA. The multifunctional Rep protein is essential for viral replication, is required early in the viral lifecycle, and functions as an amultimer, making it an ideal target for pathogen derived resistance. The rationale behind the approach was that by making transgenic maize plants constitutively express mutant Rep proteins, these over-expressed non-functional Reps would bind to any newly-synthesized viral Reps and prevent the complex from binding to the viral origin of replication. Thus, the replication of the incoming viral DNA would be inhibited, and consequently the maize plant would be resistant to MSV.

A number of different mutants were made including deletions from amino acids 180–360 ($rep^{1\text{-}179}$) and amino acids 220–360 ($rep^{1\text{-}219}$), of the 360-amino acid Rep, mutations in motif III required for virus replication (III-) and in the retinoblastoma-related protein-binding domain (RBR), a motif that is important in the viral lifecycle and is a negative regulator of the cell cycle. The final mutant used to make transgenic maize was *rep*$^{1\text{-}219Rb\text{-}}$.

Virus-Inducible Resistance

To circumvent possible negative effects on plant growth of constitutive expression of MSV-derived resistance genes, a mechanism called the "split gene cassette" has been used for inducible expression. In this case, the same mutant *rep*$^{1\text{-}219Rb\text{-}}$ gene as used above was split into two exons flanked by two viral intergenic regions containing the origin of replication and the Rep binding and nicking sites. Upon viral infection, the cassette serves as a template for rolling circle replication, during which removal of the intergenic regions results in the reconstitution of the mutant *rep*$^{1\text{-}219}$ gene (Shepherd et al. 2009).

Gene Silencing Approach

Post-transcriptional gene silencing (PTGS) occurs when translation of a targeted mRNA is prevented by translational inhibition or cleavage, with subsequent degradation of the mRNA. Given that it is a natural defence mechanism used by plants to reduce the accumulation of viral RNA, gene silencing is an attractive option for the development of MSV resistance (even though MSV is a DNA virus, there is substantial evidence that PTGS is triggered in host plants by geminivirus infection). The target chosen for this approach was a portion of the *rep* gene, chosen because of the indispensable role of the Rep protein in viral replication (Owor et al. 2011).

Testing of Genes in Model Systems for Drought Tolerance

The development of transgenic maize is impeded by the length of time required to transform this crop and the difficulty in propagating such plants in non-maize-growing regions, such as Cape Town, South Africa, without access to sophisticated computer-controlled glasshouses. We therefore developed model systems in which we could test the expression and physiological effects of the genes of interest.

Transgenic Arabidopsis

XvSap1

XvSap1 was transformed into *A. thaliana* by Ti plasmid-mediated transformation under the control of a cauliflower mosaic virus 35S promoter, a *nos* terminator, and *bar* gene (coding for resistance to bialaphos) selection using the vector pSMB. Southern blot analysis showed that transgenic plants contained one to a few copies of the gene. Western blot analysis confirmed the expression of the *XvSap1* protein (Garwe et al. 2006).

Root elongation is an accurate and convenient indicator of *Arabidopsis* seedling growth. Using this parameter, the growth of transgenic and non-transformed control plants was evaluated with plants growing on plant nutrient agar in Petri dishes, supplemented either by NaCl or mannitol. Root growth was expressed as a percentage relative to growth on unsupplemented media. There was marked growth inhibition at 100 mM NaCl in the control seedlings, whereas the transgenic lines continued to grow well, with a 43 % difference in relative growth between line 21G and the wild type.

Tolerance to osmotic stress was determined by growing *Arabidopsis* seedlings on media supplemented with mannitol. At 50 mM mannitol, both transgenic lines were clearly coping better, displaying a 47 % difference in relative growth on day seven between the transgenic line 6D and the control, and this trend continued at 100 mM mannitol.

Plants were exposed to 42 °C for 2 h and then allowed to recover at room temperature. After a week, the wild-type plants had wilted and the leaves were almost completely bleached, whereas the transgenic plants had begun to recover. At the end of 2 weeks they had new leaves and were beginning to flower, whereas the wild-type *Arabidopsis* plants were completely bleached and dead (Fig. 8.4).

XvAld

Transgenic *Arabidopsis* plants constitutively expressing *XvAld* were exposed to various abiotic stresses. Seedlings were tested in tissue culture as above. The results of exposure to mannitol (50 and 75 mM) and 14 % polyethylene glycol (PEG 8000) showed resistance to these stresses (Maredza 2007).

To evaluate stress tolerance in mature plants, dehydration stress was imposed on soil-grown plants at the beginning of the reproductive stage. After 6 days the wild-type (WT) plants were severely wilted, while the transgenic plants displayed tolerance.

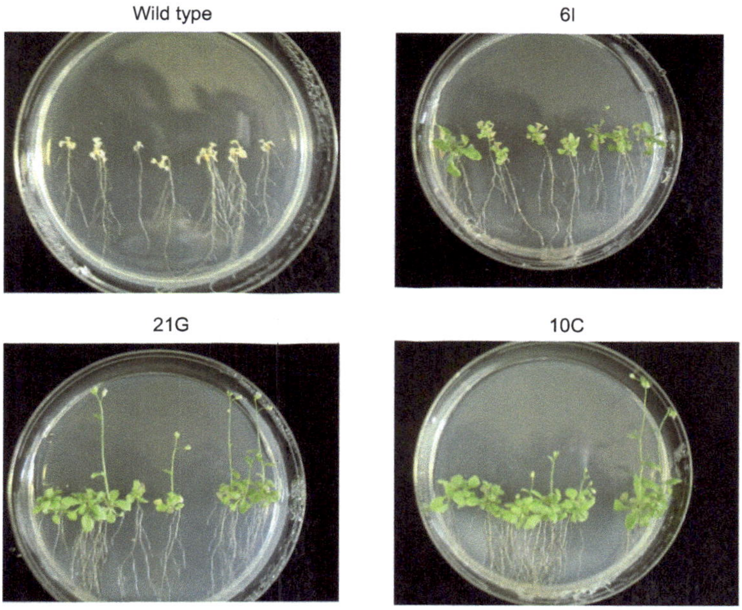

Fig. 8.4 Phenotype of *Arabidopsis* plants transformed with *XvSap1* (10C, 21G and 6I) and untransformed control plants (WT) 2 weeks after heat shock (42 °C for 2 h)

Transgenic Tobacco

Plasmid pSMB-*XvSap1* was transformed into tobacco using *Agrobacterium tumefaciens*. Southern blot analysis showed that one copy of the transgene had been integrated in each of the lines selected. Western blot analysis confirmed the expression of the *XvSap1* protein (Garwe 2003).

Transgenic and control plants were grown in media supplemented with 9 % polyethelene glycol to exert osmotic stress for 7 days. The plants were then returned to a fresh medium and photographed after 7 days of recovery. As can be seen, the two transgenic lines, A7 and A5, had recovered by this time. The control lines, on the other hand, had not.

Encouraged by these results, the decision was taken to grow tobacco plants in soil for 6 weeks, after which water was withheld. Figure 8.5 shows one transgenic plant, A7, and the wild type 10 days later. The untransformed plants became progressively chlorotic, whereas the transgenics remained green for longer.

These results, together with those obtained from transgenic *Arabidopsis* plants, encouraged us to transform both *XvSap1* and *XvAld* into maize (section "Transgenic Maize Resistant to MSV").

Fig. 8.5 Phenotype of dehydration stressed tobacco. A representative transgenic (A7) and wild type (KEI) line are shown. Photographs were taken 10 days after the imposition of the stress

Testing of Genes in Model Systems for MSV Resistance

Black Mexican Sweetcorn (BMS) in Tissue Culture

Replication of MSV DNA was assayed in BMS cells using biolistic bombardment (Shepherd et al. 2005). The ability of the various mutant Rep-expressing constructs to inhibit the replication of wild type MSV was measured in a transient assay. The effects ranged from total inhibition of replication (e.g. p*rep*$^{\text{III-Rb-}}$, p*rep*$^{\text{1-219Rb-}}$), ~80 % inhibition (p*rep*$^{\text{1-219}}$), and enhancement of replication (p*rep*$^{\text{1-179}}$), to no effect on replication. It was clear from this data that the pRBR-interaction motif mutation (Rb$^-$) contributed significantly to the protein's ability to inhibit wild-type MSV replication (Shepherd et al. 2007a, b).

Based on these results, three of the constructs that inhibited MSV replication were chosen for transformation into a model monocot plant, *Digitaria sanguinalis*.

The inducible split gene cassette containing the mutant *rep*$^{III\text{-}Rb\text{-}}$ gene was also tested in BMS. The results showed that MSV replication was greatly inhibited (there was a 25-fold reduction of viral DNA when the wild type was co-bombarded with the split gene cassette construct). Again it was decided that this data warranted the development of transgenic maize plants.

Finally, the spliceable-intron hairpin RNA (hpRNA) construct, which upon expression produces short interfering RNA (siRNA) targeted for silencing the *rep* gene, was tested in BMS. Realtime PCR analysis showed that this interfered with

MSV replication to the extent that viral DNA levels were indistinguishable from those of the negative controls. In addition, this hairpin was able to reduce the replication of diverse isolates belonging to the MSV-A group, the strain that causes severe maize streak disease (Owor et al. 2011). Based on these data, transgenic maize was made expressing the siRNA construct.

Digitaria sanguinalis

Digitaria sanguinalis is a widespread MSV-sensitive grass native to the warm, temperate sub-tropical region of Southern Africa. It was micropropagated from immature inflorescenses, and the callus transformed by particle gun bombardment. Transformants were fertile, and the time taken from bombardment to setting of seed was approximately 4 months. This proved to be an ideal model for testing genetically engineered resistance to MSV (Chen et al. 1998).

Embryogenic *D. sanguinalis* calli were transformed by particle bombardment using the mutant rep-containing constructs p*rep*$^{\text{III-}}$, p*rep*$^{\text{III-Rb-}}$, and p*rep*$^{\text{1-219Rb-}}$. Transgenic plants were challenged with MSV by using viruliferous leafhoppers (*Cicadulina mbila* Naudé). None of the calli containing *rep*$^{\text{III-}}$ regenerated. If the pRBR mutation was included, regeneration occurred, implying that interaction of the Rep$^{\text{III-}}$ protein with the host retinoblastoma-binding protein could disrupt normal cell cycle regulation. However p*rep*$^{\text{III-Rb-}}$ transgenic plants, while proving to be resistant to MSV, showed stunting and infertility. Only plants expressing Rep$^{\text{1-219Rb-}}$ were resistant to the virus, phenotypically normal and fertile. The resistance manifested itself in different plants as apparent immunity, reduced symptom severity, and decreased virus loads (Shepherd et al. 2007a, b).

Encouraged by these results, the construct carrying p*rep*$^{\text{1-219Rb–}}$ was transferred into maize (section "The Path Ahead").

The Construction of Gene Cassettes for Drought Tolerance and Transformation into African Maize Varieties

Promoters are *cis*-acting regions upstream from a gene, which determine the nature and extent of its expression. It is possible that the constitutive expression of genes involved in abiotic stress resistance could hamper the normal growth of transgenic plants (Morran et al. 2011). We therefore decided to use the stress-inducible promoter of *XvSap1* to drive the expression of the chosen *X. viscosa* genes. This step was taken because no significant similarity with any known plant promoter was identified (Odour 2009; Elick 2012).

The *XvPsap1* promoter was cloned upstream of the reporter firefly luciferase gene (*luc*) into plant vectors. As a rapid indicator of stress induction the cassette was

Fig. 8.6 Transgenic CML144 plants (*back*) and A188 (*front*) growing in the biosafety compliant glasshouse at Kenyatta University, Nairobi (Source: Miccah Seth)

transformed into black Mexican sweetcorn (BMS) in tissue culture and subjected to salt stress (200 mM NaCl). Luciferase activity peaked after 24 h. Encouraged by these results, the cassette was transformed into tobacco and maize. Expression of the gene in both transgenic tobacco and maize peaked after 3 days of dehydration.

Due to these positive data gene cassettes containing *XvPrx2*, *XvSap1*, *XvG6* and *XvAld*, driven by the *XvPsap1* promoter and followed by the *nosT* terminator from the *A. tumefaciens,* Ti plasmid were introduced in the plant vector pTF101.

Two inbred tropical CYMMIT lines CML144 and 216 were transformed and compared with the inbred A188 line that is often used in genetic transformation. Both CML lines outperformed A188 in transformation and regeneration efficiency, as well as growth of transformed lines in glasshouse conditions. CML144 was chosen for future work (Fig. 8.6).

The cassettes mentioned above are being transformed into CML 144. Preliminary results with *XvPrx2* transgenic lines which were dehydrated for 10 days, rewatered, and left to recover showed encouraging improvement compared with wild-type plants.

Open pollinated (OP) maize varieties are largely farmer-bred, providing seed that can be saved for subsequent plantings. These are different from hybrid cultivars which are specifically bred for particular traits, such as yield or stress resistance, but which have to be bought each season. Smallholder farmers often choose to forgo the benefits of hybrid maize varieties in order to plant their own seed. To date, however, no OP varieties have been improved by genetic modification. This might change in the project under discussion here, however, as scientists in Kenya and Tanzania have successfully regenerated commercial Tanzanian OP varieties (Seth et al. 2012). They found that immature zygotic embryos of 1–1.5 mm obtained at the age of 14–16 days after pollination were the best sources of ex-plant. Of the four varieties tested, Situka M-1 and Staha were found to be the most reliable.

Following this success, the gene constructs mentioned above are being transformed into Staha. Preliminary results with *XvPrx2* transgenic lines which were dehydrated for 21 days, rewatered, and left to recover, showed encouraging improvement compared to wild-type plants.

Transgenic Maize Resistant to MSV

Dominant Negative rep *Mutants*

As discussed above, although several mutant MSV *rep* constructs were shown to inhibit virus replication in transient BMS assays and in transgenic *D. sanguinalis*, only one, p*rep*$^{1\text{-}219\text{Rb-}}$, which retained amino acids 1–219 having had the C terminus deleted, and had a mutation in the pRBR binding domain, resulted in phenotypically normal, fertile, MSV-resistant plants (Shepherd et al. 2007a). The construct cloned into the plant vector pAHC25 (Christensen and Quail 1996) was transformed into maize Hi-II. The resultant plants were crossed with an elite inbred line at Pannar Seed, South Africa and tested for virus resistance in the laboratory. Resistant phenotypes included no symptom development, delayed symptoms, and mild symptoms (Shepherd et al. 2007b; Fig. 8.7).

Subsequently, new transgenics have been generated containing *rep*$^{1\text{-}219\text{Rb-}}$ in a minimal transgene cassette background. These lack selectable markers and antibiotic resistance genes. Challenges to T1 and T2 generation transgenics showed a reproducible reduction in infection percentages as well as a delay in the onset of symptoms, especially when compared to a sensitive genotype and a conventionally bred MSV-tolerant hybrid, Pan77 (Thomson et al. 2010).

The Path Ahead

The UCT MSV research group has been extremely fortunate to have had the South African seed company, Pannar, as a partner almost since the inception of the project. Not only have they provided the major funding for its development, but they have also been invaluable in assisting with the growth, crossing, and analysis

Fig. 8.7 Transgenic maize plants showing no symptoms (**a**), delayed and mild symptoms (**b** and **c**) compared with non-transgenic maize plants (**d**)

of virus resistance of transgenic plants. The value of such public/private partnerships cannot be overemphasized, and newcomers to this arena are strongly advised to investigate similar options.

The next step in bringing this crop to farmers will be field trials. Based on the outcome of such field trials, an application will be made to the Registrar of the GMO Act for commercialization. However, as outlined in Chapter 4 (Genetically Modified Crops Commercialized in South Africa), The current status of GM crops in South Africa, the outlook for the introduction of African-developed new GM crops in that country is not promising. Perhaps, as with the African Biofortified Sorghum Project, the future lies elsewhere on the continent.

The drought-resistance project is not as far advanced, and is currently in glasshouse trials at Kenyatta University. Should the data from these tests be positive, a seed company partner will be sought to carry the project further to field trials and, ultimately, commercialization. Funding for this project ends in December 2013, and as discussed by Julius Ecuru "Funding of biotechnology infrastructure without a plan for commercialization: is Africa being set up to fail?", it is critical that projects important to Africa go beyond the research phase in order to deliver products useful to farmers.

Another useful lesson can be learnt from this project: the value of regional collaboration in Africa. Maize can only be grown with difficulty in the Western Cape region of South Africa, largely due to low light intensity. We sought partners in more suitable regions of the country but found no interest. The answer came in the form of the partnership with the Department of Biochemistry and Biotechnology at Kenyatta University. The team here had the infrastructure, the know-how, but most importantly, the passion to see this project through. In agricultural biotechnology it is imperative to have good ideas, but it is essential to have perseverance as the path is long and arduous. Without passion, the chance for success is slim.

From this discussion, and despite the barriers along the way, it is clear that indigenous genes should be considered a useful source of agronomic traits for the development of transgenic crops important to Africa.

References

Bartels D, Engelhardt K, Roncarati R, Schneider K, Rotter M, Salamini F (1991) An ABA and GA modulated gene expressed in the barley embryo encodes an aldose reductase related protein. EMBO J 10:1037–1043

Bohren KM, Bullock B, Wermuth B, Gabbay KH (1989) Thealdo-detoreductase superfamily. J Biol Chem 264:9547–9551

Bradley DJ, Kjellbom P, Lamb CJ (1992) Elicitor- and wound-induced oxidative cross-linking of a proline plant cell wall protein: a novel, rapid defence response. Plant Cell 70:21–30

Castiglioni P, Warner D, Bensen RJ, Anstrom DC, Harrison J, Stoeker M, Abad M, Kumar G, Salvador S, D'Ordine R, Navarro S, Back S, Fernandes M, Targolli J, Dasgupta S, Bonin C, Luethy MH, Heard JE (2008) Bacterial RNA chaperones confer abiotic stress tolerance in plants and improved grain yield in maize under water-limited conditions. Plant Phys 147: 446–455

Chen W, Lennox SJ, Palmer KE, Thomson JA (1998) Transformation of *Digitaria sanguinalis*: a model system for testing maize streak virus resistance in Poaceae. Euphytica 104:25–31

Christensen AH, Quail PH (1996) Ubiquitin promoter-based vectors for high-level expression of selectable and/or screenable marker genes in monocotyledonous plants. Transgenic Res 5: 213–218

Dietz K-J (2011) Peroxiredoxins in plants and cyanobacteria. Antioxid Redox Signal 15: 1129–1159

Elick TL (2012) Investigation of XvSap promoters from the resurrection plant, Xerophytaviscosa. MSc thesis, University of Cape Town

Felix MM (2007) Molecular characterisation of XvRG6 and XvRF17, genes isolated from the resurrection plant Xerophytaviscosa. MSc thesis, University of Cape Town

Garwe D (2003) The characterization of *XvSap1*, a gene isolated from the resurrection plant Xerophytaviscosa Baker, and its expression in transgenic plants. Ph.D. thesis, University of Cape Town

Garwe D, Thomson JA, Mundree SG (2003) Molecular characterization of *XvSap1*, a stress-responsive gene from the resurrection plant *Xerophytaviscosa* (Baker). J Exp Bot 54:191–201

Garwe D, Thomson JA, Mundree SG (2006) *XVSAP1* from *Xerophytaviscosa* improves osmotic-, salinity- and high-temperature-stress tolerance in *Arabidopsis*. Biotechnol J 1:1137–1146

Govender K (2006) Characterisation of *XvPrx2*, a type II peroxiredoxin isolated from the resurrection plant *Xerophytaviscosa* (Baker). Ph.D. thesis, University of Cape Town

Gowrishankar J (1985) Identification of osmoresponsive genes in *Escherichia coli*: evidence for participation of potassium and proline transport systems in osmoregulation. J Bacteriol 164:434–445

Iyer R, Mundree SG, Rafudeen MS, Thomson JA (2007) Desiccation tolerance genes and avenues for drop improvement. In: Jenks M, Wood A (eds) Plant desiccation tolerance. University of Iowa Press, Iowa City

Marais S, Thomson JA, Farrant JM, Mundree SG (2005) *XvVHA- c"1-* a novel stress responsive V-ATPase subunit c" homologue isolated from the resurrection plants *Xerophytaviscosa*. Physiol Plant 122:54–61

Maredza A (2007) Isolation of the aldose reductase gene (*XvAld1*) from the resurrection plant *Xerophytaviscosa*, the characterisation of the gene product and transgenic plants expressing the gene. Ph.D. thesis, University of Cape Town

McLaggan D, Naprstek J, Buurman ET, Epstein W (1994) Interdependence of K^+ and glutamate accumulation during osmotic adaptation of *Escherichia coli*. J Biol Chem 269:1911–1917

Morran S, Eini O, Pyvovarenko T, Parent B, Singh R, Ismagul A, Eliby S, Shirley N, Langridge P, Lopato S (2011) Improvement of stress tolerance of wheat and barley by modulation of expression of DREB/CBF factors. Plant Biotechnol J 9:230–249

Mowla SB (2005) Molecular characterization of *XvPer1*, a novel antioxidant enzyme from the resurrection plant *Xerophytaviscosa*, and AC3 a LEA-like protein from *Arabidopsisthaliana*. Ph.D. thesis, University of Cape Town

Mowla SB, Thomson JA, Farrant JM, Mundree SG (2002) A novel stress-inducible antioxidant enzyme identified from the resurrection plant *Xerophytaviscosa* (Baker). Planta 215:716–726

Mundree SG (1996) Genetic determinants of salinity tolerance in tobacco. Ph.D. dissertation, Auburn University

Mundree SG, Baker B, Mowla S, Peters S, Marais S, Van der Willigen C, Govender K, Maredza A, Muyanga S, Farrant JM, Thomson JA (2002) Physiological and molecular insights into drought tolerance. Afr J Biotechnol 1:28–38

Ndima T, Farrant J, Thomson J, Mundree S (2001) Molecular characterization of XVT8, a stress-responsive gene from the resurrection plant *Xerophytaviscosa* (Baker). Plant Growth Reg 35:137–145

Odour RO (2009) Functional analyses of the novel stress-inducible *XvPsap1* promoter isolated from *Xerophytaviscosa*. Ph.D. thesis, University of Cape Town

Oliver MJ, Bewley JD (1997) Desiccation tolerance of plant tissues: a mechanistic overview. Hort Rev 18:171–214

Owor BE, Martin DP, Rybicki EP, Thomson JA, Bezuidenhout ME, Lakay FM, Shepherd DN (2011) A *rep*-based hairpin inhibits replication of diverse maize streak virus isolates in a transient assay. J Gen Virol 92:2458–2465

Peterbauer T, Mucha J, Mach L, Richter A (2002) Chain elongation of raffinose in pea seeds: isolation, characterisation, and molecular cloning of a multifunctional enzyme catalysing the synthesis of stachyose and verbascose. J Biol Chem 277:194–200

Peters S, Mundree SG, Thomson JA, Farrant JM, Keller F (2007) Protection mechanisms in resurrection plant *Xerophytaviscosa* (Baker): both sucrose and raffinose family oligosaccharides (RFOs) accumulate in leaves in response to water deficit. J Exp Bot 58:1947–1956

Serrano R, Rodriquez-Navarro A (2001) Ion homeostasis during salt stress in plants. Curr Opin Cell Biol 13:399–404

Seth MS, Bedada LT, Mneney EE, Odour RO, Machuka JS (2012) In vitro regeneration of selected commercial Tanzanian open pollinated maize varieties. Afr J Biotechnol 1:6043–6049

Shepherd DN, Martin DP, McGivern DR, Boulton MJ, Thomson JA, Rybicki EP (2005) A three-nucleotide mutation altering the Maize streak virus Rep pRBR-interaction motif reduces symptom severity in maize and partially reverts at high frequency without restoring pRBR-Rep binding. J Gen Virol 86:803–813

Shepherd DN, Mangwende T, Martin DP, Bezuidenhout M, Thomson JA, Rybicki EP (2007a) Inhibition of maize streak virus (MSV) replication by transient and transgenic expression of MSV replication-associated protein mutants. J Gen Virol 88:325–336

Shepherd DN, Mangwende T, Martin DP, Bezuidenhout M, Kloppers FJ, Carolissen CH, Monjane AL, Rybicki EP, Thomson JA (2007b) Maize streak virus-resistant transgenic maize: a first for Africa. Plant Biotechnol J 5:759–767

Shepherd DN, Martin DP, Thomson JA (2009) Transgenic strategies for developing crops resistant to geminiviruses. Plant Sci 176:1–11

Sherwin H, Farrant JM (1996) Differences in rehydration of three desiccation-tolerant angiosperm species. Ann Bot 78:703–710

Sherwin H, Farrant JM (1998) Protection mechanisms against excess light in the resurrection plants *Craterostigmawilmsii* and *Xerophytaviscosa*. Plant Growth Regul 24:203–210

Sprenger N, Keller F (2000) Allocation of raffinose family oligosaccharides to transport and storage pools in *Ajugareptans*: the roles of two distinct galactinol synthases. Plant J 21: 249–258

Taiz L (1992) The plant vacuole. J Exp Biol 172:113–122

Thomson JA, Shepherd DN, Mignouna HD (2010) Developments in agricultural biotechnology in sub-Saharan Africa. Ag Bio Forum 13(4):article 4

Chapter 9
Biotechnology of Nutritionally Enhanced Food Crops Targeting Malnutrition in Rural Agricultural Populations: The Case Study of Africa Biofortified Sorghum

Silas D. Obukosia

Abstract Biofortification of staple foods has great potential in alleviating the rampant micronutrient deficiency in low income countries, mainly in Africa, through developing staple foods with enhanced minerals and vitamins. The three most important minerals and vitamins of focus include enhanced vitamin A and bioavailable zinc and iron, while folate and iodine deficiencies are met through other approaches including fortification. Currently, on-going research in biotransformation focuses on developing four staples (rice, sorghum, cassava, and bananas) with enhanced pro-vitamin A, zinc, and iron. Of these, Africa biofortified sorghum, cassava, and bananas mainly target African countries. In the Africa Biofortified Sorghum Project, the first priority product to be commercialized will be sorghum with enhanced vitamin A, and the second priority product will be to augment pro-vitamin A with increased bioavailable zinc and iron. The primary target countries of release will be Kenya and Nigeria, representing East Africa, and West Africa Anglophone countries, respectively. Biofortification through genetic transformation complements on-going biofortification efforts through conventional breeding and fortification, especially among those rural farmers/consumers that do not buy commercial fortified foods, or consumers located in arid and semi-arid regions of Africa with little access to vegetables and fruits in the dry seasons.

Keywords Biofortification • Sorghum • Nutrition • Africa

Abbreviations

AATF	African Agricultural Technology Foundation
ABS	Africa Biofortified Sorghum

S.D. Obukosia (✉)
Africa Harvest Biotechnology Foundation International, 642-00621, NAIROBI, KENYA
e-mail: sobukosia@africaharvest.org

F. Wambugu and D. Kamanga (eds.), *Biotechnology in Africa*,
Science Policy Reports 7, DOI 10.1007/978-3-319-04001-1_9,

AHBFI	Africa Harvest Biotechnology Foundation International
ARC	Agricultural Research Council (South Africa)
ARIPO	African Regional Intellectual Property Organization
CFT	Confined field trial
CNP	Candidate novel protein
CORAF/WECARD	West and Central Council for Agriculture Research and Development
CRT I	Carotene desaturase I
CSIR	Council of Scientific and Industrial Research (South Africa)
DALY	Disability-adjusted life year
FTO	Freedom to operate
GCGH	Grand Challenge for Global Health
GI	Gastrointestinal
IAR	Institute of Agricultural Research (Nigeria)
ICRISAT	International Crops Research Institute for the Semi-Arid Tropics
INERA	Institut de l'Environnement et de Recherches Agricoles (Burkina Faso)
IPMT	Intellectual Property Management Team
KARI	Kenya Agricultural Research Institute
MAB	Marker assisted breeding
NABDA	National Biotechnology Development Agency (Nigeria)
NARS	National agricultural research system
NBA	National Biosafety Authority
NGO	Non-governmental organization
OAPI	Organisation Africaine de la Propriété intellectuelle
OPVS	Open-pollinated varieties
PMI	Phosphomannoseisomerase
PSY1	Phytoene synthase
SADC	Southern African Development Community
SGF	Simulated mammalian gastric fluid
VAD	Vitamin A deficiency
YLD	Years lived with disability
YLL	Years of life lost

Introduction

The overall goal of the Africa Biofortified Sorghum (ABS) Project is to develop and deploy highly nutritious sorghum to those farmers, consumers, and end-users in Africa that rely on sorghum as their staple food. Sorghum (*Sorghum bicolor*) is an ideal crop for many arid and semi-arid areas of Africa, because it is drought-tolerant and heat- and waterlogging-resistant. In Africa, it is a primary source of food for more than 300 million people who live mainly in the arid and semi-arid regions of the continent. Although sorghum grain has high energy content (starch content 56–73 % and 3.3 % fat), has substantial amounts of protein (11.3 % crude protein), is a rich source of B-complex vitamins and minerals (iron and phosphorus), the grain is deficient in vitamin A, the iron and zinc is not bio-available, and the protein has reduced digestibility upon cooking. Because of these deficiencies, humans with a diet that consists primarily of sorghum display vitamin A, zinc, and iron deficiencies, especially in children and mothers. Micronutrient and protein deficiencies are prevalent in Africa, especially in ABS target countries.

Importance of Micronutrients

Micronutrients that include vitamins and minerals are essential, especially in the early lives of children for robust growth and development. In particular, vitamin A, iodine, iron, zinc, and folate play pivotal roles in maintaining a healthy and productive population. Globally, at least two billion people live with vitamin and mineral deficiencies (UNICEF 2004; Anon 2009). Interventions to curb micronutrient deficiencies would contribute to Millennium Development Goals: (1) eradicate extreme poverty and hunger, (2) achieve universal primary education, (3) promote gender equality and empower women, (4) reduce child mortality; and (5) improve maternal health (Anon 2009). In May 2008, the Copenhagen Consensus panel determined that vitamin A and zinc supplementation for children provided the very best return on investment in global development from among 30 investment alternatives analyzed (Anon 2009).

Importance of Vitamin A

Vitamin A is primarily known for preventing blindness, because retinol (another name of the vitamin) is indispensable to the retina's ability to adapt to dark lighting conditions. Vitamin A also promotes healthy eye surface membranes, helping to prevent the scarring of the cornea. This makes adequate vitamin A vital for the prevention of a widespread condition called xerophthalmia, a serious eye disorder that is the primary cause of sight loss among the five million visually disabled children in the world (Whitcher et al. 2001). Additionally, vitamin A is critical to the survival and physical health of children exposed to disease, because of its ability to boost the immune system. Globally, vitamin A deficiency results in the annual death of nearly 670,000 children under 5 (Black et al. 2008), and the loss of sight of nearly 350,000 million children (Black et al. 2008; Anon 2009). Approximately one third of the world's children under the age of 5 have inadequate dietary intake of vitamin A, and are therefore ill-equipped for survival (WHO 2009). Most of the global vitamin A deficiencies in diets occur in South Asia, most of Sub-Saharan Africa, some countries in Latin America, and parts of China. For example, Kenya

Table 9.1 Micronutrient deficiency in selected African countries (Anon 2009)

Country	Pre-school age children with anaemia	Pregnant women with anaemia	Non-pregnant women with anaemia	Pre-school age children with vitamin A deficiency	Population at risk of inadequate intake of Zn
Burkina Faso	91.5 %[a]	68.3 %[a]	52.0 %	54.3%	13.3 %
Niger	81.3 %	65.5 %	62.0 %	67.0%	9.4 %
Mali	82.8	73.4	61.0	58.6	11.1
Togo	52.4	50.2	38.4	35.0	22.9
Gambia	79.4	75.1	59.1	64.0	36.1
Nigeria	76.1 %	66.7 %	62.0 %	29.5 %	12.8 %
Ghana	76.1	64.9	43.1	75.8	21.0
Kenya	69.0 %	55.1 %	46.4 %	84.4 %[b]	32.9 %
Uganda	64.1	41.2	28.7	27.9	23.8
Rwanda	41.9	10.6	59.4	6.4	39.8
South Africa	24.1 %	21.8 %	26.4 %	16.9 %	19.7 %
Zambia	52.9	46.9	29.1	54.1	38.0
Zimbabwe	19.3	18.8	34.3	35.8	38.0
Egypt	29.9 %	45.4 %	27.6 %	11.9 %	8.6 %
Sudan	84.6	57.7	43.5	27.8	10.8
Ethiopia	75.2	62.7	52.3	46.1	21.7
USA	3.1 %	5.7 %	6.9 %	0 %	9.1 %

[a]Highest proportion in the world
[b]Second highest proportion in the world

exhibits the highest rates in the world, with **84.4** % of pre-schooled children suffering from this deficiency. Other African countries show similar high levels of deficiencies—Burkina Faso (**54.3** %), Zambia (**54.1** %), Ethiopia (**46.1** %), Nigeria (**29.5** %) (Table 9.1).

Importance of Zinc

Zinc promotes immunity, resistance to infection, and the growth and development of the nervous system. It also promotes the production of antibodies against intestinal pathogens (Lazzerini et al. 2008). Diarrhoeal disease due to zinc deficiency causes 18 % of all deaths of children below 5 years of age (Bryce et al. 2005). Studies have shown that zinc supplementation, given with oral rehydration therapy, can reduce the incidence of diarrhoea in children by 27 %. It can also reduce the incidence of acute lower respiratory tract infections by 15 %. A 10–14-day course of zinc supplementation has also been shown to increase children's resistance to further episodes of diarrhoea and other disease for 2–3 months following supplementation (Anon 1999). Table 9.1 shows that Sub Saharan Africa has a very high rate of zinc deficiency. The rates of zinc deficiency, like those for vitamin A deficiency, are high across southern Africa: Kenya (**32.9** %), Burkina Faso (**13.3** %), South Africa (**19.7** %) and Nigeria (**12.8** %) (Table 9.1).

Importance of Iron

Anaemia caused by iron deficiency is responsible for the death of up to 115,000 women annually (Black et al. 2008). This accounts for one fifth of all maternal deaths (UNICEF 2008). This has the additional result of leaving tens of thousands of children without the protective care of their mothers, thus putting them at further risk of illness and death. Iron-deficiency anaemia is also estimated to cause almost 600,000 stillbirths or deaths of babies within their first week of life (Stoltzfus et al. 2004). Similar to vitamin A deficiency, anaemia prevalence is concentrated in Sub Saharan Africa, South Asia, and parts of Latin America. Table 9.1 shows a similar trend to vitamin A and Zn deficiencies in selected African countries with regard to Fe deficiencies. Burkina Faso shows the highest rates of anaemia in pre-schooled children (**91.5** %) and pregnant women (**68.3** %) in the world (Table 9.1). The ABS Project aims to eradicate micronutrient deficiency in Africa.

Potential Role of ABS in Alleviating Micronutrient Deficiencies

The root cause of mineral and vitamin deficiencies globally is low concentration of mineral and vitamins in key staples (including rice, maize, sorghum, cassava, and bananas) that constitute most of the diet of affected populations, as well inability of families to access vegetables and fruits in arid and semi-arid areas. Following an announcement by Bill Gates in 2003 for projects that could "help apply innovation in science and technology to the greatest health problems of the developing world", ABS was selected as one of the 45 projects in the Grand Challenge for Global Health (GCGH) initiative. ABS was one of the four projects in the sub-category Grand Challenge 9 (GC Number 9) program funded by the Bill and Melinda Gates Foundation with the goals of "creating a full range of optimal, bioavailable nutrients in a single, staple plant species". The four grants awarded comprised of BioCassava Plus increasing the level of vitamin A and iron; Golden Rice—High provitamin A, enhanced Fe and Zn Bioavailability; Banana 21-high vitamin A and increased iron; and the Africa Biofortified Sorghum (ABS) Project—Nutritionally Enhanced Sorghum for the Arid and Semi-Arid Tropical Areas of Africa though biofortification.

Complementary Role of Biofortification and Fortification in Alleviating Micronutrient Deficiencies

Biofortification is the development of micronutrient-dense staple crops using traditional breeding practices and modern biotechnology (Nestel et al. 2006; Pfeiffer et al. 2007). The advantages of biofortification include: the nutrients are available in staples that are consumed daily in large quantities by all family members, especially women and children who are most vulnerable (Nestel et al. 2006); the process benefits the low income households (Nestel et al. 2006); after the initial investment, recurrent investments are low, and the germplasm can be shared internationally; once in place, the biofortification system is sustainable and becomes a means of reaching malnourished people in relatively remote rural areas such as arid and semi-arid regions of Africa, where the inhabitants have limited access to commercial fortified foods and where fruits and vegetables are scarce (Nestel et al. 2006).

Fortification with iron, vitamin A and zinc averts a significant number of infant and child deaths and is a very attractive preventive health-care intervention. Fortification with iron, iodine, and potentially zinc provides significant economic benefits, and the low unit cost of food fortification ensures large benefit: cost ratios, with effects via cognition being very important for iron and iodine. Fortification, on the other hand, reaches those families who buy commercial foods, and tends to be favorable to households living in urban areas. However, fortification will not reach all individuals and is most attractive as an investment where there is a convenient food vehicle, where processing is more centralized, and where either the deficiency is widespread or the adverse effects are very costly even though only a small group is affected. Fortification requires a suitable food vehicle (Horton 2006). There are populations that are hard to reach with commercial fortification, particularly those living in more remote geographic areas and not utilizing purchased foods (Horton 2006). It is harder to reach the poorest who are the most price-sensitive and who buy lower-grade items that are less likely to be fortified, as shown clearly for iodized salt (UNICEF 2008). Biofortification and fortification are therefore highly complementary in delivery of micronutrients to the malnourished.

Cost-Effectiveness of Biofortification in Combating Micronutrient Malnutrition

The first step in assessing the cost-effectiveness of any intervention, including biofortification, is to determine the magnitude of the problem that the intervention is trying to address (Meenakshi et al. 2010). One strand of literature has focused on the productivity losses that occur as a consequence of malnutrition (Horton 1999; Horton and Ross 2003). However, 'disability-adjusted life year' (DALY) that was first detailed by Murray and Lopez (1996) is increasingly becoming a popular measure for quantifying the magnitude of ill health (Stein et al. 2005). Zimmerman and Qaim (2004) were the first to use the DALY framework in the context of biofortification. DALYs lost are the sum of years of life lost (YLL) and the years lived with disability (YLD). The YLL represents the numbers of years lost because of the preventable death of an individual, while the YLD represent the numbers of years spent in ill-health because of a preventable disease or condition.

$$\text{DALYs lost} = \text{YLL} + \text{YLD}$$

The DALYs lost from VAD are high in African countries, where 0.4–0.8 % of the population is affected. Thus, annually, 121,000 DALYs are lost to VAD in Kenya, while in Nigeria, nearly 800,000 DALYs are lost (Meenakshi et al. 2010).

The next step will be to analyze the reduction in burden of micronutrient deficiency due to the intervention. For biofortification, this will depend on development of the biofortified varieties, adoption of crop by farmers, consumption by target groups, enhanced nutrient intake, and resultant reduced DALY burden (Meenakshi et al. 2010). The final step is to analyze the cost of achieving these reductions. This cost is a factor of research and development, maintenance breeding, and dissemination.

The *ex ante* impact analysis reported that it is likely that ABS will be a "very cost-effective" health intervention. ABS has a benefit–cost ratio that is at least 10, cost per Disability Adjusted Life-Year (DALY) saved of less than $100, 20 times more cost-effective than the WHO and World Bank average. The World Development Report for 1993 (World Bank 1993), which reviewed many public health interventions, suggests that interventions costing less than $150 per DALY saved are highly cost-effective (Meenakshi et al. 2010), and interventions per capita income of $1900 are considered "very cost-effective". ABS *ex ante* social economic impact of this project showed that full implementation would result in a net present value of $139,020,241 in the next 25 years, with many other corresponding benefits to consumers.

Consortium for Implementation of the Africa Biofortified Sorghum Project

The Africa Biofortified Sorghum Project has been implemented through a consortium of 14 institutions categorized into three groups—Technology Development, Product Development, and the group for "Creating an enabling environment" (Fig. 9.1). Africa Harvest was the grantee organization in ABS Phase I, and DuPont Pioneer the lead science institution for the core partners within the ABS consortium (Anon 2013).

The Technology Development team consisted of DuPont Pioneer, Council of Scientific and Industrial Research (CSIR) of South Africa, and University of Berkeley, California. The function of the group was to conduct the scientific discovery and technology innovation that make the ABS product feasible and enhance its performance (Anon 2010b).

The Product Development institutes were The International Crops Research Institute for the Semi-Arid Tropics (ICRISAT) Kenya, Kenya Agricultural Research Institute (KARI), the Agricultural Research Council (ARC),University of Pretoria, L'Institut de l'Environnement et de Recherches Agricoles (INERA) of

Fig. 9.1 Part of Africa Biofortified Sorghum Project Consortium during the launch of Phase II, August 2010

Burkina Faso, and the Institute of Agricultural Research (IAR) of Nigeria. The function of this group was to convert the ABS technology into a product that can be distributed to farmers and consumers. Their specific functions are: the choice of sorghum varieties and introgression of ABS genes into local varieties, led by ICRISAT and NARS, and nutritional formulation studies, led by The University of Pretoria.

The Consortium for "Creating an enabling environment" included Africa Harvest Biotechnology Foundation International (AHBFI), African Agricultural Technology Foundation (AATF), West and Central Council for Agriculture Research and Development (CORAF/WECARD), and National Biotechnology Development Agency (NABDA, Nigeria). While the project was in full bloom, 84 scientists and other professional staff participated in its implementation. The function of this consortium was to provide certain capability and resources to these groups and facilitate the deployment of the product to the end user. AHBFI provided leadership for the management of the project, communication, biosafety, and regulatory systems. AATF provided expertise in IP management, while CORAF/WECARD and NABDA provided leadership in advocacy and governance.

Other ABS project consortium members that provided consultative services include Biosafety Recourse Network and Harvest Plus. They provided services in biosafety and Nutritional studies respectively. It is, however, noteworthy that the above-outlined functions of ABS Consortium were in full bloom in ABS Phase I, while in Phase II, due to limited funding, the functions have been toned down. The networks are still fully maintained through regular communication, but the number of actively participating institutions has been reduced to DuPont Pioneer, AHBFI, KARI, IAR, NABDA, and INERA.

Current Achievements of Africa Biofortified Sorghum Project

Africa Biofortified Sorghum Project has completed Phase I (July 2005–June 2012) (GCHG 2010), and is now in Phase II. The project specifically focuses on developing nutritionally-enhanced sorghum for the arid and semi-arid tropical regions of Africa.

Over time, the Project plans to develop and deploy three nutritious sorghum products in the following order: First Priority Product—sorghum with enhanced pro-vitamin A, Second Priority Product—sorghum with enhanced pro-vitamin A and bioavailable zinc and iron, Third Priority Product—sorghum with enhanced pro-vitamin A, bioavailable zinc, and iron and improved protein quality and digestability. To date, the project has achieved the following milestones:

- A sorghum transformation protocol was optimized, leading to significant increased transformation efficiencies from less than 1 % to over 10 %. This optimization of sorghum transformation capabilities provides a global opportunity for additional improvement of the sorghum crop through genetic engineering.
- The world's first "golden sorghum" transgenics were developed as a result of Phase I support. The "golden sorghum" showed enhanced levels of pro-vitamin A, reduced phytate, and an improved protein profile (Fig. 9.2). The pro-vitamin A amounts obtained from golden rice ranged up to 31.1 μg/g ß-carotene.
- Bioavailability studies have shown increased rates of zinc and iron absorption.
- The successful field and greenhouse trials have been a great source of support and encouragement. ABS has undergone over seven field trials in the USA, greenhouse trials in Kenya and South Africa, and two confined field trials in Kenya and Nigeria are complete.
- Preliminary food product formulation trials have shown that ABS can be used to successfully make a wide range of traditional African and modern food products.
- The Intellectual Property audit for freedom to operate status has been achieved for all the genes used in ABS project in all target countries and regions in Africa.
- Capacity building and infrastructural development has been done in the USA for African scientists in partnering institutions from countries of deployment in genetic transformation, throughput breeding, biosafety, and regulatory systems in readiness for Phase II.

Fig. 9.2 *Top* sorghum grains show non-transformed sorghum on the *left* and transformed on the *right*. *Bottom* is sectioned ABS showing yellow pro-vitamin A tissue

- ABS traits have been backcrossed with popular African sorghum varieties, and have shown stability in agronomic and ABS traits in African varieties (including Marcia, Gadam, Tegemeo, KARI Mtama I, Sudanese, Malisor, SAMSORG 14, 17, and 40).

Africa Biofortified Sorghum Biosafety and Regulatory Strategy

Diligent and systematic consideration of biosafety and regulatory requirements for ABS from research to commercialization is critical to the success of the ABS project. Key aspects addressed to date include compliance in the laboratory, greenhouse, and confined field trials, and food and environmental safety of ABS genes.

Regulatory Compliance for Confined Field Trails in Africa

Africa Biofortified Sorghum has undergone two seasons of confined field trials in Kenya and Nigeria in 2011 and 2012, while an application was submitted to the National Biosafety Authority (NBA) for greenhouse testing approval. The latter is currently waiting on the completion of a Biosafety Level II greenhouse facility. The objectives of the CFT trials were threefold: to introgress ABS pro-vitamin A traits in local sorghum cultivars, study the stability of ABS traits over generations, and develop biosafety data on the impact of ABS gene flow on the environment. Two categories of biosafety measures were stipulated: genetic and material confinement. Five categories of methods and procedures undertaken by the ABS Project to enforce genetic isolation were:

(i) Using an isolation distance of at least 400 m from the nearest cross-compatible sorghum species;
(ii) Bagging the sorghum panicles 1 week before the reproductive phase
(iii) Growing ABS sorghum under bird netting to keep birds from eating sorghum seeds
(iv) Post-harvest monitoring of the experimental site for 6 months, bearing in mind the self-pollinating nature of sorghum
(v) Deflowering before anthesis.

Five material confined measures used included: Use of chain link fence reinforced by an electrical fence, 24/7 security, material incineration, and record of all persons/materials entering and or leaving the storehouse. These confinement measures have proven adequate for continued ABS CFT product development in Kenya (Figs. 9.3 and 9.4) and Nigeria.

Biosafety of Africa Biofortified Sorghum Genes

Africa Biofortified Sorghum contains two genes used in carotenoid pathway genes—*psy1* that encode phytoene synthase (PSY1), and *crtl* that encode carotene desaturase I (CRT I) to catalyze the biosynthesis of pro -vitamin A. The *pmi* gene which encodes phosphomannoseisomerase (PMI) has been used as the selectable marker gene. The biosafety studies of ABS genes are yet to be done. However, genes similar to ABS have been used in the development of Golden Rice, from which food safety information could be inferred as we await studies of the latter.

Source of genes: The vitamin A enhancing genes inserted in ABS were isolated from naturally occurring organisms that are already widespread and prevalent in the environment as constituents of food and feed (maize, sorghum, and peas). The other genes, *crtl* and *pm I* have been isolated from the soil bacterium *Erwinia uredovora* and a non-pathogenic strain of *E. coli* respectively. The *pmi* gene, on the other hand, has been obtained from maize. Extensive toxicity and allergenicity tests has been

Fig. 9.3 Confined field trial of Africa Biofortified Sorghum in Kenya (2012)

Fig. 9.4 KARI Mtama I in confined field trial in Kenya prior to backcrossing to Africa Biofortified Sorghum

carried out on proteins obtained from similar genes and used during the development of the Golden Rice project. Results have shown that these gene products were neither toxic nor allergy-inducing. These studies have been based on bioinformatics analysis followed by Simulated Gastric Fluid digestion.

Allergenicity studies: A bioinformatics analysis of PMI, PSYI, and CRT I transgene proteins showed that the protein products are not allergenic to humans (Goodman and Wise 2006). In this analysis, two sequence alignments and similarity scoring algorithms were used. A FASTA3 algorithm was used to evaluate overall alignment of each query sequence compared to all sequences in allergen online, identifying matches of low *E* scores values and or greater than 50 % identity as an indication of potential cross-reactivity. FASTA 3 was also used to search for 80 or more amino acids that aligned with a match of 35 % identity or more compared to any sequence in Allergen Online; this was suggested as a lower limit for considering potential cross-reactivity (Codex 2003). Finally, BLASTP was used to identify any significant similarity to any newly reported allergen sequences not found in Allergen Online version 6. None of the results from the three bioinformatics analyses (i.e., FASTA3 Allergen Online Full length search, FASTA3 Allergen Online by 80 amino acid segments, BLAST OF NCBI Entrez *allergen*) with all the vitamin A proteins and the selectable marker gene *pmi* met the criteria that might suggest potential allergenicity or cross-reactivity (Goodman and Wise 2006).

To further support the above findings, PMI, PSYI, and CRT I gene products were subjected to simulated mammalian gastric fluid (SGF). SGF is routinely used to assess the in-vitro digestibility of the candidate novel proteins (CNPs). Proteins that are unstable in the GI system are more likely to be safe following oral consumption than those that resist digestion, if for no other reason than they are unlikely to retain biological activity following degradation. The results showed high digestibility of PMI, PSYI and CRT I, further substantiating the results from bioinformatics analysis.

Toxicological studies: Similarly, results from bioinformatics analysis of PMI, PSY and CRT I for toxicity have shown that they are safe. Additionally, a digestibility assay has also been carried out for Golden Rice, and the results from the study showed that the expressed genes were not toxic (Goodman and Wise 2006). Also, there are no reports of toxicity to plants and animals in crops where similar modifications have been done (such as Golden Rice); in fact, some transgenic plants containing *pmi* as the selectable marker have been deregulated for food and feed (Golden Rice 2009; Bruce et al. 2008; Chassy 2010).

Feeding Studies. Feeding trials with human adults in China were carried out to measure the effect of dietary fat on bioconversion and bioavailability (Astwood et al. 1996; Kimber and Dearman 2002; Goodman and Wise 2006; Golden Rice 2009). ABS intends to undertake similar studies with the pro-vitamin A gene products from genes inserted in sorghum, but has for now taken results obtained from studies utilizing Golden Rice.

Addressing Gene Flow Concerns

One of the concerns in deployment of ABS traits was the impact of ABS gene flow on the environment. Following a request by ABS External Advisory Board, a panel of experts was constituted, and has provided its views on sorghum with respect to gene flow. The panel consisted of six members: Barb Schall, Spencer T. Olin (Professor of Biology, Washington University); Norm Ellstrand (Professor of Genetics, University of California, Riverside); Jeff Pederson (Research Geneticist, USDA/ARS, Lincoln, Nebraska); Alan Raybould (Scientific Fellow, Syngenta Corporation, United Kingdom); Prof. Patrick Ayiecho Olweny (Geneticist/Plant Breeder, Member of Parliament in Kenya) and Dr. Jeremy Ouedraogo (Geneticist; Member of Parliament in Burkina Faso) (Anon 2010c). The key findings were:

(i) Gene flow between cultivated and wild sorghum occurs with some frequency. Neutral genes from cultivated sorghum such as the nutritional genes used in developing ABS are not expected to have a selective advantage in the wild.
(ii) Gene flow from crop plants to wild relatives or landraces has resulted mainly in an increase in genetic diversity. It is therefore not expected that gene flow from ABS sorghum into wild sorghum or landraces will alter the genetic diversity any differently than gene flow from other sorghum varieties.
(iii) Environmental impacts commonly of regulatory concern, including yield loss in crops due to increases in pest pressure or weediness, or loss of diversity in flora or fauna due to invasiveness or toxicity, are not likely to occur due to the presence of ABS genes in cultivated or wild sorghum.
(iv) A thorough characterization of the transgenic plant compared to the non-transgenic plant, for agronomic performance, fitness-related characteristics, toxicity, or nutritional composition, would demonstrate that there have been no significant unintended changes and support the assessment that negative environmental impacts following gene flow from ABS sorghum are not likely.
(v) A study to compare fitness-related characteristics in 'ABS x wild' hybrids and 'non-ABS x wild' hybrids would provide evidence that unexpected gene-interactions will not significantly alter the weediness or invasiveness of hybrids. This comparison would provide additional confidence that negative environmental impacts related to gene flow are not likely. The experiments on the impact of gene flow when ABS is crossed with wild relatives are ongoing both in Kenya and in Nebraska University (Anon 2010c).

Regional Approach to Deployment of ABS in Africa

Kenya was chosen as the primary site for deregulating and deployment of ABS products for Eastern Africa, while in Western Africa the project is focused on Nigeria (8.028 million tonnes/year) and Burkina Faso (which produces 1.4 million tonnes of sorghum annually). Secondary countries for product deployment will be

South Africa (0.293 million tonnes/year) and Egypt (0.945 million tonnes/year) (Deb et al. 2004). The project activities in primary countries of product deployment are supported by in-country enthusiasm for the project, with Kenya and Burkina Faso already having enacted Biosafety Acts, while in Nigeria the Biosafety Bill has undergone several readings in the parliament. These three countries are sites of deployment to reach our goal of impacting at least 30 million people with our product and creating the potential to reach 300 million who eat sorghum in Africa. ABS project intends to accelerate trait introgression into local farmer popular varieties and hybrids by identifying varieties in common use in target countries representative of the sub-regions. The initial trait introgression will be conducted in Kenya using MAB technology. Once the varieties are converted, they will be transferred back to the originating countries, thereby avoiding delays that may ensue due to the lack of biosafety laws in those countries.

Choice of Sorghum Parent for Introgression

To ensure large-scale impact of the ABS technology on the livelihoods of people, ABS traits will be incorporated into sorghum varieties that enjoy wide regional adaptation and acceptance. These varieties were developed by a joint effort of ICRISAT and NARS. Two primary regions for selected for deployment include West Africa—Francophone and Anglo-phone—and Southern Africa and Northern Africa. SAMSORG 14, 17, and 40 were selected for the West Africa Anglophone zone. SAMSORG 40 (ICSV 400) is a short-season variety (matures in 95–100 days) adapted to Sudan savannah ecology. It yields 2.5–3.5 tonnes per hectare. SAMSORG 17 (KSV3 (SK5912) is a long-season variety (matures in 165–175 days) adapted to Southern Guinea ecology. It yields 2.5–3.5 tonnes per hectare. SAMSORG 17 (KSV3 (SK5912) is a long season variety (matures in 165–175 days) adapted to Southern Guinea ecology. It yields 2.5–3.5 tonnes per hectare. SAMSORG 14 (KSV8) is a medium-season variety (matures in 130–140 days) adapted to Northern Guinea savannah ecology (Ogbonna 2008).

For Eastern and Southern Africa, ABS traits will be introgressed into Tegemeo (2KX 17/B/I), Macia (SDS 3220), KARI Mtama I, and Gadam varieties of sorghum. The improved sorghum variety Macia (SDS 3220) was released on 14 December 1999 by the Tanzania National Variety Release Committee. Macia is a high-yielding, early-maturing, white-grained variety developed jointly by ICRISAT and national scientists in southern Africa. It has so far been released in five SADC countries—Mozambique, Botswana (under the name Phofu), Zimbabwe, Namibia, and Tanzania. It is suitable for areas with a growing season of 3–4 months, and yields up to 4 tonnes per hectare (ICRISAT 2000; Monyo et al 2004; Saadan et al. 2000) (Fig. 9.5).

KARI Mtama I sorghum variety is mainly grown in Kenya, matures in 3–3.5 months, and has white grains and the potential for wide adaptation. It grows in moist mid- latitudes of Kenya (Busia, Siaya, Homa Bay), semi-arid low lands

Fig. 9.5 SAMSORG 17, under confined field trial at the Institute for Agricultural Research Zaria, Kaduna

(Kitui, Makuweni, Mwingi, Ntharaka) and humid Coast (Kwale, Kilifi, TaitaTaveta) (Anon 2013). The tegemeo sorghum variety was released in Tanzania in 1986, and on-station yields of 4.2 tonnes per hectares have been reported there, with a yield advantage of 114 % over the local unimproved cultivars. It is also grown in Kenya.

Addressing Intellectual Property Concerns

Iintellectual property (IP) constraints are often perceived as barriers to market entry, especially when it comes to developing countries (Anon 2010a; Anon 2007). The ABS Intellectual Property Management Team (IPMG) led by Africa Agricultural Technology Forum (AATF) emulated the example of the Golden Rice Project, and negotiated ABS IP upfront. IPMG conducted and inventoried all technologies—genes, promoters, and associated genetic materials and related IP being used or scheduled to be used in the project. The freedom to operate (FTO) exercise was conducted at the point when the project was initiated, and IP updates are conducted regularly to ensure eventual commercialization is unaffected. The gene donation from partners whose input is critical are outlined below-

- Selectable marker gene—from Syngenta
- Super binary vector transformation system—from Japan Tobacco Co.
- Sorghum transformation technology—from DuPont Pioneer
- Nutritious genes used—these are owned by DuPont Pioneer
- *psy* and *crt* genes from Freiburg University in Germany/Syngenta

The Africa Biofortified Sorghum report confirms no IP impediments to the freedom to operate and use transgenic sorghum in Africa. Further, the report states that the ABS project may be executed and used in the 16 countries of the African Regional Intellectual Property Organization (ARIPO) and 16 countries of l'Organisation Africaine de la Propriété intellectuelle (OAPI) without infringing on the IP of third parties (Anon 2007).

Earlier Adoptions of Improved Technology

For biofortified foods to have an impact on the nutritional status of rural households, they require widespread adoption of the technology by farmers, acceptance of the product by consumers, presence of a functional market system for the product, and added economic value for the end-user (Nestel et al. 2006). Several strategies have been put in place/will be in place to facilitate widespread adoption of ABS, including introgression of nutritional traits into sorghum varieties with wide regional adaptation and farmers' acceptance, food formulation studies to ensure the suitability of ABS in currently consumed sorghum dishes and products, a communication strategy emphasizing the nutritional benefits of ABS, developing a seed system to ensure availability of pure ABS seeds to farmers, linking ABS to nutritional programs targeting schools, hospitals and food aid for the vulnerable communities, and plans for the future breeding of ABS nutrition traits into sorghum hybrids, in addition to OPVS to improve on farmer agricultural productivity.

The successful adoption of the orange sweet potato, and its impact on the nutritional status of consumers in Uganda, Kenya, and Mozambique, portend potential success for future biofortification programs (Haskell et al. 2004; Hotz et al. 2012; Jalal et al. 1998; van Jaarsveld et al. 2005; Low et al. 2007). Additionally, the relatively high adoption of improved sorghum varieties in target areas in Africa implies that when the right facilitative measures are put in place, the same high adoption could be realized for ABS. Nestel et al. (2006) proposed some of these measures, which include use of farmer participatory breeding methods to identify the locally adapted biofortified genotypes that best suit producer–consumer needs, ensuring good access to planting material through the development of seed systems, and the development of markets for both the harvested biofortified crop (s) and any processed products made from them, such as complementary foods. Consequently, significant assistance will be needed to determine, understand, and identify the actions needed to be undertaken in order to overcome constraints on farmer adoption.

Capacity-Building

One of the pertinent components of successful and sustainable technology transfer is building on local capacity to carry out the technologies deployed. The ABS consortium showcases a North–South and South–South collaboration in technology development and transfer. One key output of this network was the building of local capacity to support the three components of the consortium. DuPont Pioneer played the major role in capacity building by training twelve Research Fellows in technology, ranging from sorghum transformation to throughput sorghum breeding. The trainees were drawn from CSIR, KARI, INERA, and ARC from Kenya, Burkina Faso, and South Africa. Additionally, 20 scientists from KARI, INERA, and IAR were trained to carry out greenhouse and confined field trial experimentation of ABS. There is also regular on-going training as required by the National Biosafety Regulation systems to train all personnel working in the sorghum confined field trials. The latter is undertaken in Nigeria and Kenya, before CFT planting and harvesting. Other South-based activity included training two scientists, from Nigeria and Kenya, in the use of molecular markers in plant breeding at ARC, South Africa.

Conclusion

The biofortification of sorghum has unmatched potential to impact the nutritional status of sorghum consumers in Africa, not only in the rural areas but also the hard-to-reach arid and semi-arid areas with few alternative crop choices. The choice of sorghum as a crop—highly adapted to harsh climatic conditions, its diverse uses, and the imminent impacts of global warming—puts ABS in an unprecedented position to create an appreciable impact on nutrition and food security in the targeted region.

The successful transformation of sorghum and the heightening of its efficiency opens up the crop for further improvement. The successful regulatory compliance for confined field trials in Africa and the appropriate communication support in Africa, the availability of ICRISAT/NARS varieties for trait introgression, and the diverse and competent capacity built suggest that ABS product development can proceed without impediments on the path to product deployment.

The fact that the ABS project has Golden Rice as it forerunner has also accrued many advantages and leveraging that greatly contribute to ABS's current and future success. To mention a few such advantages: since similar genes are used in both Golden Rice and sorghum, intellectual property issues have been pro-actively addressed, and issues such as optimum nutritional serving of the grain have been addressed/or are being addressed due to the organization's familiarity with the earlier crop.

Perhaps one of the biggest assets of the ABS Project was the establishment of a multi-disciplinary consortium comprised of diverse institutions, including the private sector, NARs, universities, NGOs, and regional organizations. The Project also called for a wide range of roles for technology development and product development, creating an enabling environment with the ability to effectively develop and deploy the ABS product. This team brings in a multi-disciplinary approach that is pertinent to successful product development and deployment.

The chief constraint of ABS, currently, is inconsistent funding. The project greatly appreciates the funding received from the Bill and Melinda Gates Foundation (which financed Phase I) and the Howard Buffet Foundation (which is funding on-going Phase II activities). However, for successful product development and deployment, there is still a need for more funding. It would be especially convenient if a multi-donor funding approach were adopted.

Acknowledgements This is to thank the ABS Consortium Members and institutions for their invaluable contribution to the achievements of the Project. We are grateful to the Bill and Melinda Gates Foundation for funding ABS Phase I, and the on-going Howard Buffet Foundation funding of ABS Phase II. Our gratitude is also extended to the institutions that DONATED (not gave) ABS IP rights, including Syngenta, DuPont Pioneer, Japan Tobacco, and Freiburg University in Germany/Syngenta.

References

Anon (1999) Zinc Investigators' Collaborative Group. Prevention of diarrhoea and pneumonia by zinc supplementation in children in developing countries: pooled analysis of randomized controlled trials. J Pediatr 135:689–697

Anon (2007) Intellectual property management. A global vision with an African focus to fight poor nutrition with nutrient-rich crops. Africa Biofortified Sorghum Project. Mid-term Report December 2007. http://www.africaharvest.org

Anon (2009) United call to action on vitamin and mineral deficiencies. Global Report 2009. http://www.unitedcallaction.org/documents/Investing_in_the_future.pdf

Anon (2010a) Intellectual property management. Africa Biofortified Sorghum Project. Five-year Progress Report, p 82. http://www.biosorghum.org. Accessed March 2013

Anon (2010b) Innovative partnerships for accelerated technology development; the case study of Africa Biofortified Sorghum. Africa Biofortified Sorghum Project. Five-year Progress Report, pp 36–37. http://www.biosorghum.org. Accessed March 2013

Anon (2010c) Comprehensive strategy to address gene flow issues. Africa Biofortified Sorghum Project. Five-year Progress Report, pp 78–81. http://www.biosorghum.org. Accessed March 2013

Anon (2013) Sorghum: general information and agronomic aspects (www.infonet-bioviion.org/print/ct/127/crops)

Astwood JD, Leach JN, Fuchs RL (1996) Stability of food allergens to digestion in vitro. Nat Biotechnol 14:1269–1272

Black RE, Allen LH, Bhutta ZA, Caulfield LE, de Onis M, Ezzati M, Mathers C, Rivera J (2008) Maternal and child undernutrition: global and regional exposures and health consequences. Lancet 371(9608):16. http://www.who-int/nutrition/topics/lancetseries_undernutrition1.pdf. Accessed 25 March 2013

Bruce C, Egnin M, Gao Y, Glenn K, Kleter A, Nestel P, Newell-McGloughlin M, Shillito R (2008) Nutritional and safety assessments of foods and feeds nutritionally improved through biotechnology: case studies. Compr Rev Food Sci Feed Saf 7:R131–R137
Bryce J, Boschi-Pinto C, Shibuya K, Black RE (2005) WHO Child Health Epidemiology Reference Group. WHO estimates of the causes of death in children. Lancet 365(9465):1147–1152
Chassy B (2010) Food safety risks and consumer health. New Biotechnol 27(5):534–544
Codex Alimentarius Commission (2003) Alinorm 03/34: Joint FAO/WHO Food Standard Programme, Codex Alimentarius Commission, Twenty-Fifth session, Rome, Italy 30 June–5 July 2003. Appendix III, Guideline for the conduct of food safety assessment of foods derived from the recombinant-DNA plants and Appendix IV, Annex on the assessment of possible allergenicity, pp 47–60
Deb UK, Bantilan MCS, Roy AD, Rao PP (2004) Global sorghum production scenario. In: Bantlan MCS, Deb UK, Gowda CLL, Reddy BVS, Obilana AB, Evenson RE (eds) Sorghum genetic enhancement—research process, dissemination, and impact. International Crops Research Institute for the Semi-Arid Tropics, Patancheru
GCHG (2010) Grand challenge in global health—the grand challenges: goals. Available at http://www.grandchallenges.org/Pages/BrowseByGoal.aspx. Accessed 15 Jan 2010
Golden Rice (2009) Research on biosafety. http://www.goldenrice.org/Content2-How/how_bio safety.html. Accessed March 2010
Goodman RE, Wise J (2006) Bioinformatics analysis of proteins in Golden Rice 2 to access potential allergenic cross-reactivity. Food Allergy Research and Resource Program University of Nebraska. Study No 02-2006, pp 1–24
Haskell MJ, Jamil KM, Hassan F, Peerson JM, Hossain MI, Fuchs GJ, Brown KH (2004) Daily consumption of Indian spinach (*Basella alba*) or sweet potatoes has a positive effect on total-body vitamin A stores in Bangladeshi men. Am J Clin Nutr 80:705–714
Horton S (1999) Opportunities for investments in nutrition in low-income Asia. Asian Dev Rev 17 (1–2):246–273
Horton S (2006) The economics of food fortification. Am Soc Nutr 136(4):1068–1071
Hotz C, Loechl C, Lubowa A, Tumwine JK, Ndeezi G, Masawi AN, Baingana R, Carriquiry A, de Brauw A, Meenakshi JV, Gilligan DO (2012) Introduction of b-Carotene-rich orange sweet potato in rural Uganda resulted in increased vitamin A intakes among children and women and improved vitamin A status among children 1–3. Am J Clin Nutr 142:1871–1880
Horton S, Ross J (2003) The economics of iron deficiency. Food Policy 28:51–75
ICRISAT (2000) Sorghum variety Macia released in Tanzania. Int Sorghum Millets Newsl 41:7, ISSN 1023–487X
Jalal F, Nesheim MC, Agus Z, Sanjur D, Habicht JP (1998) Serum retinol concentrations in children are affected by food sources of beta-carotene, fat intake, and anthelmintic drug treatment. Am J Clin Nutr 68:623–629
Kimber I, Dearman RJ (2002) Approaches to assement of allergenic potential of novel proteins in food derived from genetically modified crops. Toxicol Sci 68:4–8
Lazzerini M, Ronfani L (2008) Oral zinc for treating diarrhoea in children. Cochrane Database Syst Rev (3):1–67. http://www.thecohranelibrary.com. Accessed 19 Feb 2014
Low JW, Arimond M, Osman N, Dunguara B, Zano F, Tschirley D (2007) Orange-fleshed sweet potatoes increased vitamin A intake and serum retinol concentrations in young children in rural Mozambique. J Nutr 137:1320–1327
Meenakshi JV, Nancy LJ, Victor MM, Hugo D, Josyline J, David Y, Firdousi N, Carolina G, James G, Erika M (2010) How cost-effective is biofortification in combating micronutrient malnutrition? An ex ante assessment. World Devel 38(1):64–75
Monyo ES, Ngereza J, Mgonja MA, Rohrbach DD, Sadan HM, Ngowi P (2004) Adaptation of improved sorghum and pearl millet technologies in Tanzania. http://www.icrisat.org/publica tions/EbooksOnline/Publications-2004/2004_adoption1.pdf
Murray CJL, Lopez AD (1996) The global burden of disease. Cambridge University Press, Washington, DC

Nestel P, Bouis HE, Meenakshi JV, Pfeiffer W (2006) Biofortification of staple food crops. J Nutr 136:1064–1067. http://jn.nutrition.org/content/136/4/1064.full.pdf+html. Accessed 19 February 2014

Ogbonna AC (2008) Sorghum and environmentally friendly industrial crop in Nigeria. http://wwwnyen.sik.se/traditionalgrains/review/Poster%20Presentations%20PDF%20files/Ogbonna.pdf. Accessed 19 Feb 2014

Pfeiffer WH, McClafferty B (2007) HarvestPlus: breeding crops for better nutrition. Crop Sci 47 (Suppl 3):S88–S105

Saadan HM, Mgonja MA, Obilana AB (2000) Performance of sorghum variety Macia in multiple environments in Tanzania. http://oar.icrisat.org/1897/1/ismn41_10–12_2000

Stein A, Meenakshi JV, Qaim M, Nestel P, Sachdev HPS, Bhutta Z (2005) Analyzing the health benefits of biofortified staple crops by means of the disability-adjusted life years approach: a handbook focusing on iron, zinc, and vitamin A. HarvestPlus technical monograph 4. International Food Policy Research Institute (IFPRI) and International Center for Tropical Agriculture (CIAT), Washington DC/Cali

Stoltzfus RJ, Mullany L, Black RE (2004) Iron deficiency anaemia. In: Ezzati M et al (eds) Comparative quantification of health risks: global and regional burden of diseases attributable to selected major risk factors. World Health Organization, Geneva, p 164

UNICEF (2004) Vitamin and mineral deficiency: a global progress report. UNICEF, New York. Available from http://www.micronutrient.org/reports/default.asp. Accessed 25 March 2013

UNICEF (2008) Progress for children: a report card on maternal mortality. UNICEF, New York, p 4

van Jaarsveld PJ, Faber M, Tanumihardjo SA, Nestel P, Lombard CJ, Spinnler Benadé AJ (2005) β-Carotene-rich orange-fleshed sweet potato improves the vitamin A status of primary school children assessed with the modified-relative-dose-response test. Am J Clin Nutr 81:1080

Whitcher JP, Srinivasan M, Upadhyay MP (2001) Corneal blindness: a global perspective. Bull World Health Organ 79(3):217

WHO (2009) Global prevalence of vitamin A deficiency in populations at risk 1995–2005. WHO global database on vitamin A deficiency. World Health Organization, Geneva

World Bank (1993) World Development Report 1993. World Bank, Washington, DC

Zimmerman R, Qaim M (2004) Potential health benefits of Golden Rice: a Philippine case study. Food Policy 29:147–168

Part IV
Capacity Development

Chapter 10
The Role of African Universities in Training and Mentorship of Biotechnology Scientists to Embark on Future Challenges in Africa

Idah Sithole-Niang

Abstract This paper reviews the role of African Universities in postgraduate training of biotechnology scientists in agriculture. It highlights examples of various graduate programs that have been established, and the impact they have had so far. The paper also highlights the role and impact of various development partners. It also discusses various initiatives in agricultural biotechnology that have been established on the African continent, and their impact on capacity-building. It concludes by showcasing the Masters degree program in Biotechnology at the University of Zimbabwe, which trained 82 postgraduate students, over 2 decades, and is currently being revived.

Keywords Agricultural biotechnology • Capacity building • MSc Biotechnology Degree Programme • AU/NEPAD Centers of Excellence • Biotechnology training and mentorship

Abbreviations

AATF	African Agricultural Technology Foundation
ABI	African Biotechnology Initiative
ABSPII	Agricultural Biotechnology Support Project II
ACCI	African Center for Crop Improvement (South Africa)
AGERI	Agricultural Genetic Engineering Research Institute (Egypt)
AGRA	Alliance for a Green Revolution in Africa
ARC	Agricultural Research Council (South Africa)
ASBMB	American Society for Biochemistry and Molecular Biology

I. Sithole-Niang (✉)
University of Zimbabwe, Department of Biochemistry, PO MP 167, Mt. Pleasant, Harare, Zimbabwe
e-mail: isniangzw@yahoo.com

F. Wambugu and D. Kamanga (eds.), *Biotechnology in Africa*, Science Policy Reports 7, DOI 10.1007/978-3-319-04001-1_10,

AU/NEPAD/ ABNE	African Union African Biosafety Network of Expertise (Burkina Faso)
BeCA	Biosciences East and Central Africa (Kenya)
Bio FISA	Finnish Southern Africa Partnership Programme to Strengthen NEPAD/SANBio Network
BMGF	Bill and Melinda Gates Foundation
BNARI	Biotechnology and Nuclear Agriculture Research Institute (Ghana)
BTZ	Biotechnology Trust of Zimbabwe
CARP	Community Action Research Program
CBSV	Cassava brown streak virus
CFT	Confined field trial
CIAT	International Center for Tropical Agriculture
CMV	Cassava African mosaic virus
COP	Conference of Parties
CSIR	Council for Scientific and Industrial Research
CYMMIT	International Maize and Wheat Improvement Centre
DDPSC	Donald Danforth Plant Science Center
DFID	UK Department for International Development
DST	Department of Science and Technology (South Africa)
FARA	Forum for Agricultural Research in Africa
GTZ	Gesellschaft für Technische Zusammenarbeit
HGBF	Howard G. Buffet Foundation
IAR	Institute for Agricultural Research (Nigeria)
ICGEB	International Center for Genetic Engineering and Biotechnology (South Africa)
IFPRI/PBS	International Food Policy Research Institute Program for Biosafety Systems
IITA	International Institute of Tropical Agriculture
ILRI	International Livestock Research Institute (Kenya)
IPBO	Institute for Plant Biotechnology Outreach (Belgium)
ISAAA	International Service for the Acquisition of Agri-biotech Applications
KARI	Kenya Agricultural Research Institute
MARI	Mikocheni Agricultural Research Institute (Tanzania)
MOP	Meeting of the Parties
MSU	Michigan State University
NABNET	North Africa Biosciences Network (Egypt)
NARO	National Agricultural Research Organisation (Uganda)
NARS	National Agricultural Research Systems
NEPAD	African Union New Partnership for Africa's Development
NEWEST	Nitrogen-use efficiency, water-use efficiency and salt tolerance
NGICA	Network for the Genetic Improvement of Cowpea for Africa
NRCRI	National Root Crops Research Institute (Nigeria)

PBS	Program for Biosafety Systems
PRRI	Public Research and Regulations Initiative
QUT	Queensland University of Technology
RAEIN–Africa	Regional Agriculture and Environment Initiatives Network–Africa
RF	Replicative form
RUFORUM	Regional Universities Forum for Capacity Building in Agriculture
SABIMA	Strengthening Capacity for Safe Biotechnology Management in sub-Saharan Africa
SADC	Southern African Development Community
SANBio	Southern African Network for Biosciences
SANGL	Southern African GM Detection Laboratories
SCARDA	Strengthening Capacity for Agricultural Research and Development in Africa
SFSD	Sustainable Farming Systems Database
SRO	Sub-regional organization
TWAS	The World Academy of the Sciences
UCB	University of California Berkeley
USAID	United States Agency for International Development
WABNET	West African Biosciences Network (Senegal)
WACCI	West African Crop Improvement Center (Ghana)
WARDA	West Africa Rice Development Association (Benin)
ZIMBAC	Zimbabwe Biotechnology Advisory Council

Introduction

Very few universities in Africa provide training in agricultural biotechnology. Consequently, the type of training that is ongoing has to be defined from the outset. Training in agricultural biotechnology has taken many forms. There are students who are trained through the normal degree programs offered at universities, with some of the students being trained locally and some undergoing sandwich programs with universities in the industrialized countries, or in some cases, universities in the developing countries. Good examples of universities in the developing countries that actively participate in this process can be found in South Africa, Brazil, and India. South Africa has gone a step further, and facilitated this process by allowing students from the Southern African Development Community (SADC) to pay the same fees as their local students. However, in some professional disciplines such as medicine, there has been an agreement signed among SADC countries that allows students to attend medical school in one country, but upon completion, they have to go back to their own countries and pursue their residency, and then can choose to leave thereafter.

Table 10.1 Advanced GM Technologies in Africa as of 2012

Country	Crop	Trait under testing	Stage	Partners
Uganda	Maize	Drought tolerance	CFT, 2nd season	NARO, AATF, Monsanto, BMGF, and HGBF
Uganda	Banana	Bacterial wilt resistance	CFT	NARO, AATF, IITA
Uganda	Banana	Nutritional enhancement (Fe and pro-vitamin A)	CFT	NARO, QUT
Uganda	Cassava	Virus resistance	CFT, 2nd season	NARO, DDPSC, IITA
Uganda	Cotton	Bollworm resistance and herbicide tolerance	CFT, 3rd season	NARO
South Africa	Maize	Drought tolerant		AATF, ARC, Monsanto, BMGF, and HGBF
South Africa	Cassava	Biofortified and modified starch		HarvestPlus
South Africa	Sugarcane	Virus resistance, increased yields, alternative products		South Africa
South Africa	Maize	Maize IR resistant to MSV		University of Cape Town, Pannar Seed Co.
South Africa	Potatoes	Insect resistance		ARC and MSU
Burkina Faso	Cowpea	Insect resistance	CFT	AATF, NGICA, IITA, CSIR, Monsanto
Egypt	Maize (*Zea mays*)	Insect resistance	CFT	Pioneer, AGERI
Egypt	Cotton (*Gossypium barbadense*)	Insect resistant	CFTs	ARC
Egypt	Wheat (*Triticum durum* L.)	Drought tolerant	CFTs	AGERI
Egypt	Wheat (*Triticum durum* L.)	Fungal resistance	CFTs	AGERI
Egypt	Wheat (*Triticum durum* L.)	Fungal resistance	CFTs	AGERI
Egypt	Potato (*Solanum tuberosum* L.)	Viral resistance	CFTs	AGERI
Kenya	Maize (*Zea mays*)	Insect resistance (Insect-Resistant Maize for Africa against stem borers)	CFTs	KARI, CIMMYT, Monsanto, University of Ottawa, SFSD

(continued)

Table 10.1 (continued)

Country	Crop	Trait under testing	Stage	Partners
Kenya	Maize (*Zea mays*)	Drought tolerance (WEMA)	CFTs, 2nd season	AATF, CIMMYT, KARI, Monsanto, BMGF, and HGBF
Kenya	Cotton, (*Gossypium hirsutum* L.)	Insect resistance (bollworms)	CFTs complete	KARI/Monsanto
Kenya	Cassava (*Manihot esculenta*)	Disease resistance (cassava mosaic viral disease)	CFT, 1st season	KARI, DDPSC
Kenya	BioCassava Plus	BioCassava Plus, enhanced levels of iron and zinc, protein, Vitamin A and E	CFT, 1st season	DDPSC, KARI, IITA, CIAT
Nigeria	Cassava (*Manihot esculenta*)	Increased level of beta-carotene (pro-vitamin A)	CFT, 3rd season	DDPSC, NRCRI
Nigeria	Cassava (*Manihot esculenta*)	Nutrition enhancement for increase in iron level	CFT, 2nd season	DDPSC, NRCRI
Nigeria	Cowpea (*Vigna unguiculata*)	Insect resistance	CFT, 3rd season	AATF, NGICA, IITA, Purdue University; Monsanto, RF, USAID, DFID, CSIR, INERA, The Kirkhouse Trust, IAR
Nigeria	Sorghum (*Sorghum bicolor*)	Bioavailability of iron, zinc, protein, Vitamin A	CFT	Africa Harvest, Pioneer Hi-Bred International Inc.; CSIR, ICRISAT; AATF, FARA University of Pretoria, ARC; UCB, IAR

Source: Jose Falck-Zepeda (2012)

At the same time, some of the National Agricultural Research Systems (NARS) also undertake training activities in order to spearhead their breeding programs. When scientists are trained inhouse, as is the case for the NARS, on projects, their training is more relevant, and finds immediate application. Scientists develop confidence in their work place that cannot be equated to graduate students at a university where their training might be in a vacuum. Interestingly, in Uganda and Tanzania, some of the well-equipped laboratories and indeed some of the significant research outputs can be traced to the NARS (Table 10.1). Internationally, one can find relevant training at centers such as the International Center for Genetic Engineering and Biotechnology (ICGEB) in both Trieste and India, and the Institute for Plant Biotechnology Outreach (IPBO) in Ghent University, Belgium (Sopory 2011; Gheysen 2011). These are world-class institutions that have hosted scientists from developing countries. The World Academy of the Sciences (TWAS) has also been instrumental in facilitating funding for short-term training and fostering south–south collaboration.

The Role of Universities

On the African continent, several models are being rolled out. One model that was recently initiated is the Community Action Research Program (CARP) under the Regional Universities Forum for Capacity Building in Agriculture (RUFORUM). It involves three universities, namely Makerere University in Uganda, Moi University in Kenya, and Bunda College of Agriculture in Malawi. The objective of this program is to "*improve the relevance and effectiveness of agricultural education at African universities*". This visionary approach came from a gap analysis study conducted by RUFORUM in 2010, under the banner of *"Shifting from Outreach to engagement: transforming Universities' response to current development trends in agricultural research and training in Eastern, Central and Southern Africa"*. In an effort to address agricultural relevance to African smallholder farmers, the RUFORUM has put forward yet another initiative through a regional Ph.D. program that facilitates agricultural innovation processes in Africa. This second model, under a collaborative agreement between three African Universities (Makerere University, Egerton University, and Sokoine University of Agriculture) and two European universities (Wageningen University and Montpellier SupAgro), is funded by the ACP-EU-Edulink Program.

The Role of Centers of Excellence

Centers of excellence also provide avenues of training. They do so by '*increasing access to affordable, world-class research facilities and strengthening human resources in biosciences and related disciplines in Africa*'. A case in point are the four African Union New Partnership for Africa's Development (AU/NEPAD) Centers of Excellence, namely:

- The Biosciences East and Central Africa (BeCA), based in Nairobi at the International Livestock Research Institute (ILRI) Campus
- The Southern African Network for Biosciences (SANBio), based at the Council for Scientific and Industrial Research (CSIR), Tshwane, South Africa
- West African Biosciences Network (WABNET), based in Dakar, Senegal, for the West African countries
- North Africa Biosciences Network (NABNET), based in Cairo, Egypt, for countries north of the Sahara

The BeCA in particular is involved in training and co-supervision of thesis research, conducting both short courses and traveling seminars throughout the eastern and central African sub-region. Highlighting the achievements at these centers is an example closer to home (FARA, 2012).

In his opening address in a speech delivered on his behalf by his Special Assistant, Mr. Abdoul Salaam Bello, His Excellency Dr. Ibrahim Assane Miyaki, the CEO of NEPAD Planning and Coordinating Agency, re-affirmed the state of institutional capacity in Africa to being that where a capacity is lacking in one country, it could in essence be substituted by a capacity existing in a neighboring country or region. As such, a country working strategically with its neighbours could, in essence, have its needs met that way while a much longer term strategy is being put in place. This is easier said than done when linkages and partnership do not exist. As such, one of NEPAD's key thrusts has been that of establishing such networks and partnerships. Some of the key projects accomplished are under one of their Centers of Excellence, namely the SANBio, and specifically the BioFISA program. Through the BioFISA program, and with funding from the Department of Science and Technology (DST) and the government of Finland, the just-ended Phase I saw eight projects being assisted in order to build both infrastructural and human capacity as well as bring products to markets (Mumba 2012). As key examples, the fisheries project saw pond fishing facilities being installed and aquaculture skills being instilled in small-scale farmers in Dowa district in Malawi; 600 small-scale farmers from three countries—Malawi, Namibia, and Swaziland—and an additional 100 agricultural extension officers from other countries being trained in transferring the skills to other countries. Overall, there were 30 doctoral and masters students who were trained on the eight BioFISA projects during its lifetime of 4 years (Mumba 2012). Going forward, BioFISA Phase II will assume a slightly different approach, one that will mirror the primary objectives of enunciated by African Heads of State in their African Biotechnology Initiative (ABI) (Silfverberg 2012; Juma and Serageldin 2007).

The problem of not conducting appropriate training for agricultural biotechnology is not unique. A recent report by the American Society for Biochemistry and Molecular Biology (ASBMB) observed that there was a *"huge disconnect between how we currently train scientists and the actual employment opportunities available for them"* (Rosenberg 2012). Other sectors also face the same problems. Universities are largely ignorant of the needs of the agricultural biotech industry. There is a tendency to self-perpetuate, as lecturers tend to teach what they know. It is not unusual to find lecturers who still work in the same area of expertise as they did when they were doctoral students. There is a tendency to have a "business as usual" approach to teaching, where lecturers do not want to venture outside their area of expertise and begin to address the needs of the industry they purportedly serve. Rosenberg reckons there is an apparent reluctance on the part of the science education system to proactively seek what industry needs so that this can be passed on to their students. Universities perceive changing the curriculum as a lot of work. A case in point is this author's university, which had to fast-track the need to introduce honours across all its undergraduate degree programs in order to compete with the new and upcoming universities within the country. Just to illustrate a point, when the author joined the university 2 decades ago, he once remarked at a departmental board meeting that his undergraduate biochemistry degree program in the UK offered honours to all the students in his class. The response was that the honours option was only offered to a few students who had averaged an upper second class pass in all the ten subjects under review. Eighteen years later, the student numbers dwindled; in the 20th year when the honours progam was introduced, record numbers of students signed up for the first year! What's more, they came to the department with excellent grades, whereas in the past the department would mostly get students who did not qualify for any of the more popular programs such as medicine, pharmacy, medical laboratory sciences, and food science.

Traditionally, universities play a key role in education, training, and mentorship of scientists. To get away from the business as usual approach, universities must engage their local stakeholders and solicit input on the type of graduate they should produce to better serve the community. There needs to be proactive engagement on the part of the universities themselves. Sometimes this deliberate re-arrangement is brought about by donors and development partners who, by virtue of their funding, will "force" certain multidisciplinary collaborations to occur. These can take the form of regional institutions collaborating together, or they might involve advanced institutions in the north. When students are trained in advanced laboratories, the same donors can create a mechanism to ensure that the trainees return to their home institutions. Similarly, the home institution should also have a mechanism of ensuring their staff development fellows return home. In fact, this arrangement must be so attractive that the fellows are eager to return. In some cases there have been fellowships that ensure that upon completion of their studies or visits, the fellows receive additional support for equipment and start-up funds. Previously, the Rockefeller Foundation established Biotechnology Career fellowships that enabled fellows to visit an advanced laboratory for 3 months each year for 3 years. This fellowship morphed into another arrangement that involved the fellows receiving research funding for two additional years following completion of their studies. Nowadays, the foundation, through their Alliance for a Green Revolution in Africa (AGRA) doctoral program for plant breeders, have also established a loan facility for their fellows, following completion of their studies either at the African Center for Crop Improvement (ACCI) at the University of Kwa Zulu–Natal or the West African Crop Improvement Center (WACCI) at the University of Ghana, Legon, to assist with the establishment of seed companies (Table 10.2).

1.4

The role of the National Agricultural Research Systems (NARS): these days, the NARS also play a key role in training and staff development. In Uganda, the Kawanda Agricultural Research Institute is now a world leader in banana transformation, having conducted numerous confined field trials for biofortified banana with provitamin A, zinc, and iron, banana with a cell cycle regulatory gene for rapid

Table 10.2 Universities, Research Institutes and Centres of Excellence involved in teaching and practice of agricultural biotechnology in 12 select African countries

Name of research institution	Location
A. Universities	
University of Development Studies	Ghana
University of Ghana Legon (WACCI)	Ghana
Kenyatta University	Kenya
Jomo Kenyatta University of Agriculture and Technology	Kenya
University of Nairobi	Kenya
Bunda College of Agriculture	Malawi
University of Cape Town	South Africa
University of Pretoria	South Africa
University of the Witwatersrand	South Africa
University of Kwa Zulu Natal (ACCI)	South Africa
Makerere University	Uganda
University of Dar es Salaam	Tanzania
Sokoine University of Agriculture	Tanzania
University of Zimbabwe	Zimbabwe
National University of Science and Technology	Zimbabwe
B. Research institutes	
Institut de l'Environnement et de Recherches Agricoles (INERA)	Burkina Faso
BNARI	Ghana
KARI	Kenya
BeCA	Kenya
Agricultural Research Council	South Africa
CSIR/SANBio	South Africa
Kawanda Research Institute, NARO	Uganda
Namulonge Research Institute, NARO	Uganda
Mikocheni Research Institute	Tanzania
Tobacco Research Board	Zimbabwe
C. International and regional centers	
AHBFI	Kenya, South Africa, USA
AATF	Kenya, Nigeria, Uganda, South Africa
ABNE	Burkina Faso, Uganda
ABSPII	Uganda
CIAT	Mozambique, Uganda
CIMMYT	Kenya, Zimbabwe
FARA/SABIMA/SCARDA	Ghana, SADC, ASARECA, WECARD
IITA	Nigeria, Uganda, Tanzania
ICGEB	South Africa
ICRISAT	Kenya, Zimbabwe
IFPRI/PBS	Kenya, Uganda, Malawi, Nigeria
RAEIN-Africa	Namibia
WARDA	Benin

Source: Sithole-Niang (2011b)

growth, and banana with bacterial wilt or weevil resistance (Sithole-Niang 2011a). Work of equal magnitude is also ongoing at the sister institution at Namulonge Research Institute in cassava crop improvement, and also a prime location for CFT and an insectary for Bt maize and cotton. In Tanzania, at Mikocheni Agricultural Research Institute (MARI), a laboratory for cassava transformation has been established with funding from the Bill and Melinda Gates Foundation (BMGF) (Ndunguru 2012). The transformation laboratory is a biosafety level 2 containment facility for transgenic work centered two problematic viruses, Cassava African mosaic virus (CMV) and Cassava brown streak virus (CBSV). Under the leadership of Dr. Nduguru, researchers from seven NARS—Kenya, Malawi, Mozambique, Rwanda, Tanzania, Uganda, and Zambia—received training in sample collection, data analysis, and the polymerase chain reaction technique and its various applications as well as disease mapping. Diagnostic laboratories have been established in all seven countries, and are fully equipped and operational. While this is a welcome move on the part of scientists in Tanzania, especially in showing such leadership in science, the lack of an enabling biosafety regulatory framework is likely to have negative consequences for rapid developments in this area.

The Role of Development Partners

The role of development partners and how they shape the agricultural biotechnology landscape cannot be overemphasized. A complete treatise of the entire development landscape is beyond the scope of this paper. However, notable examples include such partners as the United States Agency for International Development (USAID) through their funding of the Program for Biosafety Systems (PBS), a consortium of biosafety expertise on the continent and in Asia, the Agricultural Biotechnology Support Project II (ABSPII) project, and the African Agricultural Technology Foundation (AATF) through its projects with other partners, such as the Bt Cowpea Project for *Maruca vitrata* resistance or the Nitrogen-use Efficiency, Water-use Efficiency and Salt Tolerance (NEWEST) rice project, to mention just a few. The Bill and Melinda Gates Foundation changed the face of agricultural biotechnology funding by being the first to allocate significant funding to an agricultural project as never seen before. This funding was awarded to the Africa Biofortified Sorghum Project through Africa Harvest and its various partners (Table 10.1). The foundation has gone on to fund key biosafety projects in Africa, including the African Union African Biosafety Network of Expertise (AU/NEPAD/ABNE) located in Burkina Faso, as well as the International Center for Genetic Engineering and Biotechnology (ICGEB) at the University of Cape Town, South Africa. Of note here is that there is another Biosafety Program within the AU funded by the Gesellschaft für Technische Zusammenarbeit (GTZ) which seems to work at cross-purposes with the rest of the like-minded initiatives such as PBS, ABNE, ISAAA Africenter, ICGEB, and Public Research and Regulations Initiative (PRRI). Working at cross-purposes has created a dilemma on the continent where previously the AU Heads of States had come out unambiguously and supported biotechnology development, and now there seems be ambivalence and no coherent voice on the subject (Juma and Serageldin 2007). This has a knock-on effect, in that no meaningful investment will come to the sector as long as there is status quo. Juma goes on to observe that some of these issues continue to beset us simply because there is a lack of strategic advice being given to leaders in Africa. He further notes that to date, there is no African leader who has a chief scientific advisor (Juma 2012). Clearly, sitting and doing nothing is no longer an option, but making a deliberate attempt at appointing an advisor would be a step in the right direction. This can take various forms; one could use the services of an individual expert or indeed, a whole slew of experts, or even cast the net wider to tap into the expertise in the African diaspora.

A recent survey conducted on the status of agricultural biotechnology in six Strengthening Capacity for Safe Biotechnology Management in sub-Saharan Africa (SABIMA)-participating countries (Burkina Faso, Ghana, Kenya, Malawi, Nigeria, and Uganda) hinted at the dilemma of failing to fulfil the training needs of these countries (FARA 2009). The study recommended that FARA, together with its sub-regional organizations (SROs), should be tasked with establishing a Biotechnology Cooperation Service that would link African institutions to genetic engineering facilities throughout the world. The study also highlighted the dilemma of some scientists totally lacking the background required for their jobs. One mechanism suggested for the AU/NEPAD, FARA, and its SROs was to facilitate the appointment of world-class scientists as professorial chairs at some key universities in Africa, or tap into the skills of Africans in the diaspora, so that training could be localized. In essence, the skills could be tapped at two levels—one at advisory level and another at the training and mentorship level.

The M.Sc. Biotechnology Program at the University of Zimbabwe

The M.Sc. Biotechnology Program was started in 1991, and was funded by two donors—Sida/Sarec and the Directorate-General for International Cooperation (DGIS), the Dutch Ministry of Foreign Affairs. For the following 2 years and for the rest of the program (1993–2007), funding was solely by the Dutch government. The funds were made available to what was formerly known as the Zimbabwe Biotechnology Advisory Council (ZIMBAC) and later transformed into the Biotechnology Trust of Zimbabwe (BTZ). The BTZ remains relocated in Zimbabwe and has a functional board, but with minimal activities to date. The Dutch government, however, has continued funding the Regional Agriculture and Environment Initiatives Network–Africa (RAEIN–Africa), which is now located at the University of Namibia in Windhoek, Namibia. Within the region, RAEIN–Africa has been involved in biosafety training and capacity building to enhance the negotiation skills of African Regulators who attend the Conference of Parties (COP)/Meeting of the Parties (MOP) (COP/MOP) meetings and indeed other international fora. They have a keen interest in GMO testing, and have initiated the Southern African GM Detection Laboratories (SANGL) network to build capacity in GMO-detection in participating countries. These days they, together with the International Food Policy Research Institute Program for Biosafety Systems (IFPRI/PBS), are working on the guidelines for socio-economic risk assessment.

The RAEIN–Africa, however, did not continue overseeing the facilitation of the M.Sc. Program as in-kind support was realized through teaching assistantships offered by the University of Zimbabwe. In 2007, by the end of the program, the University of Zimbabwe had trained 82 graduates, and these graduates are gainfully employed at local, regional, and international research institutions and industries (Sithole-Niang 2011b). This is one example on the continent where the local expertise actually achieved this. This was a competitive program that was fully funded, with additional funding earmarked for the immunology and virology sections, including participation in sandwich programmes at Wageningen University of Agriculture. Additionally, the technical staff also received capacity-building in order to support the running of practical classes as well as maintenance of equipment. Further, this program was actually a regional M.Sc. program attracting students from as far as Cameroon, Kenya, Uganda, Botswana, Zambia, and South Africa. To date, there is still no bonafide M.Sc. Biotechnology program that is run on a regional basis. Interestingly, however, Eduardo Mondlane University in Mozambique also launched an M.Sc. Biotechnology program in 2011. But in this progam, enrolment has largely been of students from lusophone-speaking countries, presumably due to limited numbers of students from the region who are fluent in Portuguese. News from the University of Zimbabwe indicates that the MSc Biotechnology Program is scheduled to begin in February 2013 (Sunday Mail 2012). This is a welcome move on the part of the students, as there clearly is still a niche for such a program.

Conclusions

Governments in Africa must pledge to invest 10 % of GDP into agriculture. They must be proactive and be informed of past and present experience, and be willing to learn from others. The case of the GMO debate is one such example, where African governments could enquire as to why there is wide-spread adoption of GM technology in the Americas but not on their own home soil. The training that is on-going within projects at NARS is to be encouraged, as also the role that philanthropic organizations are playing to change the scientific landscape of the African continent. The achievements of AU/NEPAD through its Centers of Excellence should be applauded, and its role in advising African Heads of States given more publicity. Lastly, it is regrettable to note that the situation in Africa largely remains fragile, as much of the progress recorded can be attributed to a large extent to individuals (some close to retirement) rather than to institutions.

References

Falck-Zepeda J (2012) The GM crop regulatory pipeline in Africa—within the next five years? http://socioeconomicbiosafety.wordpress.com//?s=The+GM+crop+regulatory+pipeline+&search=Go

FARA (2009) Status of biotechnology and biosafety in sub-Saharan Africa. Forum for Agricultural Research in Africa, Accra

FARA (2012) Africa in search of safe and high-quality biotech crops. 1st Pan-African conference on Stewardship of agricultural biotechnology, 29–30 November 2011. Forum for Agricultural Research in Africa, Accra

Gheysen G (2011) Training and education in biotechnology and biosafety. Enhancing human capacities: training and education. Food and Agriculture Organization of the United Nations, Guadalajara

Juma C (2012) Global food for thought: biotechnology and Africa's strategic interests. The Chicago Council on Global Affairs. http://globalfoodforthought.typepad.com/global-food-for-thought/2012/12/commentary-biotechnology-and-africas-strategic-interests.html

Juma C, Serageldin I (2007) Freedom to innovate: biotechnology in Africa's development. African Union, New Partnership for Africa's Development, Addis Ababa

Mumba L (2012) Southern African Biosciences Network Newsletter. Special Edition.

Ndunguru J (2012) Status of biotechnology in Tanzania. The PBS-COSTECH Tanzania Country Planning Meeting, 14–15 Aug 2012, White Sands Hotel, Dar es Salaam, Tanzania

Rosenberg M (2012) An honourable career in academia vs an alternative career in the private sector. American Society for Biochemistry and Molecular Biology. ASBMB Today 12–13 Aug

Sithole-Niang I (2011a) Transgenic horticultural crops on the African continent. In: Scorza R, Mou B (eds) Transgenic horticultural crops: challenges and opportunities. CRC Press, Boca Raton, pp 147–154

Sithole-Niang I (2011b) Teaching agricultural biotechnology and biosafety: some lessons from Zimbabwe and the Program for Biosafety Systems. Enhancing human capacities: training and education. Food and Agriculture Organization of the United Nations, Rome

Sopory S (2011) Enhancing human capacities: training and education. Food and Agriculture Organization of the United Nations, Rome

Silfverberg P (2012) BioFISA II planning workshop proceedings, 6 Sept 2012. Irene Country Lodge, Pretoria

Sunday Mail, 28 Oct 2012. M.Sc. Biotechnology Advert

Part V
Biotechnology and Biosafety Regulation

Chapter 11
Elements of an Enabling Biosafety and Regulatory Environment

Muffy Koch

Abstract Regardless of whether countries use existing laws to regulate genetically modified organisms, or develop stand-alone biosafety Acts with regulations and guidelines, the existence of legislation alone does not ensure a workable or an enabling biosafety framework. Elements essential to implementing an enabling biosafety system include clear policy objectives, the political will to make decisions, coordinated inter-ministerial decision making, efficient and transparent decision making, and all the aspects needed to implement a functional biosafety framework. These include iterative consultation with applicants, science-based risk assessment, risk management that is commensurate with the identified level of risk, a cost-effective process, and fast-tracking procedures for activities known to have low risk. Functional regional biosafety processes could facilitate coordinated decision-making across many countries, which would increase the access farmers and end users have to better planting materials and other improved products of genetic modification. In addition to reviewing the key elements for an enabling biosafety and regulatory environment, this chapter reviews historical reasons for the high number of unworkable national biosafety frameworks in Africa, and provides links to useful resources for countries wishing to revise their biosafety systems in order to enable access to the benefits of biotechnology in Africa.

Keywords Risk assessment • Biosafety communication • Biosafety law • Functional biosafety

Abbreviations

COMESA Common Market for Eastern and Southern Africa
ECOWAS Economic Community Of West African States

M. Koch (✉)
Global Biosafety Consultant, 4502 Donnelly Drive, RR4, Merrickville, ON K0G 1N0, Canada
e-mail: muffykoch@globalbiosafety.ca

F. Wambugu and D. Kamanga (eds.), *Biotechnology in Africa*,
Science Policy Reports 7, DOI 10.1007/978-3-319-04001-1_11,

GMO	Genetically modified organism
UNEP/ GEF	United Nations Environment Programme/Global Environment Facility

Introduction

Past publications that have addressed aspects required for an enabling and workable biosafety and regulatory system for genetically modified organisms (GMOs) have approached this from different perspectives. Traynor et al. (2002) outlined terms of reference for national biosafety committees, and key requirements for effective reviews and decision making. In the same year, Morris and Koch (2002) identified key differences between biosafety regulations in Africa and those in the developed world, which included the consideration of benefits, socioeconomic impact, and public input in decision making. Morris and Koch suggested mechanisms to ensure efficient implementation of African biosafety systems. McLean et al. (2002) reported on a conceptual framework for implementing biosafety that identified five key elements—national policy on biosafety, an inventory and evaluation of national legislation, biosafety capacity base for implementation, the development of regulations, and their implementation. These elements were presented to provide guidance to countries participating in the UNEP/GEF Global Project on the Development of National Biosafety Frameworks.

In spite of this abundance of practical experience, the African countries accessing funding under the UNEP/GEF process leaned heavily on the African Model Law on Safety in Biotechnology (African Union 2001), which was developed by consultants with limited genetic modification knowledge or regulatory experience, and with the intention of stalling the introduction of biotechnology. The publication of an alternative workable Model Act by the biotechnology industry (Abramson and Van der Meer 2002) was largely ignored by developing countries using the UNEP/GEF process, even though the Model Act contained provisions for a transparent, effective, and workable national biosafety regulatory framework and was completely compliant with the Cartagena Protocol on Biosafety.

Early identification of the pitfalls in the flawed AU Model Law did little to prevent its use. Despite the AU's acknowledgement of the flaws, a more recent update of the Model Law has entrenched the same problems (African Union 2011). The result was approximately 37 African countries with draft regulations and policies that were unworkable and were impossible to implement without an overhaul. Table 11.1 provides a summary of the status of biosafety frameworks in Africa. One might think that the activists who infiltrated the AU achieved their goal, but in reality, adoption of biotechnology has not slowed down, with many countries regulating and using biotechnology products (James 2012). However, Africa's access to the benefits is still threatened, and it may be years before end users on the continent benefit from approved new products.

Table 11.1 Status of national biosafety frameworks in African countries, November 2013

Status	Country
Functioning	*Policy in place*
	Burkina Faso, Egypt, Kenya, Namibia, South Africa
	Policy awaiting completion
	Ghana, Nigeria, Zimbabwe, Uganda
Being revised	Mozambique, Tanzania, Malawi
Non-functioning	Algeria, Benin, Botswana, Burundi, Cameroon, Cape Verde, Central African Republic, Chad, Comoros, Congo, DR Congo, Cote d'Ivoire, Djibouti, Eritrea, Ethiopia, Gabon, Gambia, Guinea, Guinea Bissau, Lesotho, Liberia, Libya, Madagascar, Mali, Mauritania, Mauritius, Morocco, Niger, Rwanda, Sao Tome & Principe, Senegal, Seychelles, Sierra Leone, Sudan, Swaziland, Togo, Tunisia, Zambia
No framework	Angola, Somalia, South Sudan, Western Sahara

More recently, in an attempt to address unworkable national biosafety frameworks in Africa, delegates at a 2011 workshop in Kigali, Rwanda, reached consensus on a document drafted by a legal team that outlined a 'fit-for-purpose' regulatory system, which addressed key policy and decision-making criteria (ICGEB 2011).

This chapter outlines key elements of policy, regulations, and implementation that will help to provide an enabling and efficient biosafety system for products of biotechnology.

Policy and Laws

Political will is a key element to implementing national biosafety regulations in African countries. Even after extensive consultation on draft bills, it is time-consuming to get a new bill passed through the legislature and signed by the head of state. Once signed, there are frequently additional delays in drafting and approving regulations and during the political nomination and appointment of regulatory staff members to set up the regulatory structures. In South Africa, the government took 2 years to implement the new Act, and the time frame has been even longer in other countries such as Ghana and Nigeria.

National policy on the adoption of biotechnology needs to state clearly the principles that guide the use of the technology, and the goals of the biosafety system, including what is acceptable risk. The biosafety policy should harmonise with policy for national imperatives such as sustainable development, food security, human health, job creation, and environmental protection. This should be the context for decision making. Clarity in the policy should guide decisive and effective decisions on the approval of safe and responsible biotechnology products. A summary of key elements of an enabling policy for biosafety include (ICGEB 2011):

- Recognition of the country's approach to the development and use of GMOs
- Clearly defined decision-making responsibilities
- The level of participation by other government ministries, departments, and agencies
- Goals that decision making should address with respect to safety and access to benefits
- Guidance on the use of information from other country review processes
- Guidance on sharing technical capacity with other countries
- The requirement to build technical capacity in biosafety

Inter-ministerial Co-ordination

Co-ordinating the regulatory mandates of various government ministries, agencies, and departments is another key element for ensuring a functional biosafety system. Biotechnology tools are effective across many sectors, with applications in agriculture, health, environment, forestry, and industry, all of which have existing, workable regulatory frameworks. Although the UNEP/GEF process encouraged a careful review of existing legal instruments, many of the African countries chose to develop a stand-alone law to regulate the safety of GMOs and, following the African Model Law, did little to harmonize the proposed biosafety processes and requirements with existing regulations. Instead, these countries favoured the establishment of large biosafety committees composed of representatives from a wide range of potential stakeholders, including other regulatory agencies. The number and seniority of these committee members, together with the expense of gathering them for meetings, makes it nearly impossible to call sufficient meetings to implement an effective biosafety process.

Countries that have implemented their biosafety systems, such as Burkina Faso and Ghana, have scaled down the size of their decision-making and advisory bodies to make them functional, and have established other mechanisms to ensure co-ordination between existing regulatory agencies such as quarantine control, food safety, animal health, and import and export control. A co-ordinated approach to biosafety regulation with harmonisation between government agencies is a key element in establishing a workable biosafety process.

Decision Making

Even with a structurally sound biosafety policy and process, the ability to make timely decisions founded on scientific evidence is not always possible. The South African biosafety system was established after 10 years of experience with the regulation of GMOs under an interim biosafety process. This experience enabled policy makers to address all the key requirements for a functional biosafety system, and yet, after passing the Genetically Modified Organisms Act (Act 15, 1997), the efficiency and transparency of the biosafety process dropped considerably and 15 years later, decisions are still unacceptably slow, show a weak science base, and, in some cases, are barely comprehensible.

Science-based decision making is essential for biosafety. The safety recommendations delivered by scientific advisory bodies need to be structured, thorough, and conclusive. They should highlight the potential risks that have been identified, and indicate what risk management measures are required to ensure that the risk level of the activity is acceptable for local release environments. Decision-makers need to accept and use these technical recommendations for their determinations.

For lower risk activities such as contained use (laboratory, greenhouse, and screenhouse work), confined use (field trials and clinical trials) and transit (movement of a GMO shipment through a country to a neighbouring country), the risk assessment recommendations are generally sufficient for decision-making. Activities that release GMOs into the environment for commercialization, planting, distribution to end users, and human consumption will need a safety assessment that reviews both environmental and food and feed safety.

Science-Based Risk Assessment

With considerable focus on the identification of possible risk, there is sometimes a failure to complete the risk assessment process by assessing the likelihood and the consequences of the identified hazard actually occurring. Without this, decision makers sit with lists of potential hazards, and no clear understanding of how and if these will result in significant harm. Even when likelihood is high and harm may result, there is still the possibility of applying risk management terms and conditions to reduce the identified risks to an acceptable level.

In an attempt to rationalise information requirements for risk assessment, the industry has developed a problem formulation method to identify key risks and address these with biosafety research and risk management (Raybould 2006). Regulators benefit from this refined assessment of risk, but also need to address the concerns raised by local communities, even if these have been filtered out by the problem formulation method as not relevant to the GMO in question. Addressing public concerns is an important tool for raising public understanding of GMOs and the biosafety process.

Commensurate Risk Management

While risk assessment enables regulators to make sound decisions in the absence of complete information, it is important that the risk management levels should reflect the identified risks of specific activities with GMOs. If risk is low, then risk management measures should also be low, saving extremely stringent terms and conditions for real risk that could result in significant, irreparable harm. Regulators need to ensure that in setting commensurate risk management conditions, they have taken into consideration the ability to manage the risks and to reverse unintended harm, should it occur. Setting unnecessary management measures on low-risk activities sets precedents that are difficult to reverse when familiarity and experience raise the comfort level with the GMOs in question.

Risk Communication

There are two aspects to risk communication that need to be addressed by functioning regulatory frameworks: information on how biosafety is implemented, and information on what decisions have been made on activities with GMOs in the country. Both sets of information help to raise public awareness of the role of biosafety and the approval of GMOs.

Stakeholders need clear guidelines on the scope of the regulations and what actions they must take to get approval for regulated activities with GMOs. Interested members of the public will want to know how safety is assessed, and who is responsible for decision-making. They will also want to know when there are opportunities for public participation, and how to submit comments and concerns. A functioning biosafety regulatory system should have an up-to-date communication strategy that includes timelines for key communication actions and an emergency communication protocol to deal with unintended releases, non-compliance issues, and challenges to the regulatory process.

Socioeconomic Issues

Many African countries encourage the consideration of socioeconomic issues as part of decision-making. These considerations include religious and ethical issues, and impact on vulnerable communities or trade. Importantly, socioeconomic issues relate to applications for general release of GMOs and do not have a bearing on contained or confined use, which involve short-term projects with limited impact on communities. Socioeconomic issues should be considered separately from the biosafety risk assessment and should include benefits, although this was largely overlooked in the UNEP/GEF development process.

Notably, while safety risk assessments are science-based and structured, socioeconomic issues that have been used to support decisions on food and feed import and general release are frequently based on hearsay or perception. Both Namibia and South Africa have made biosafety decisions on unsubstantiated socioeconomic claims from affected stakeholders, only to have to reverse these decisions once evidence proved them to be unfounded. The onus is on the regulatory decision-makers to ensure that claims of significant, negative socioeconomic impact are substantiated with verifiable evidence, as is their requirement for safety claims.

Iterative Consultation

Regulatory systems without mechanisms for iterative consultation between the regulators and applicants result in processing inefficiency and unnecessary delays in decision making. An iterative process enables the regulators to consult with applicants for clarity on issues until they are satisfied that they have all the information they need for decision making. Culturally, in Africa, there has been concern that direct and on-going contact between regulatory officers and applicants may open a window for corruption and manipulation, but transparency in iterative communication, with checks and balances, is easy to achieve.

Cost Effective Biosafety

In an attempt to be all-inclusive, many African biosafety regulatory systems are too bulky to be efficient or affordable. A small decision-making body of three to six people who are knowledgeable, empowered, and able to meet easily and regularly is an essential element for a workable biosafety process. Add to this a small core group of scientists who are able to contract required technical expertise as needed, an electronic document handling system, cost-effective communication protocols, and the judicious use of public hearings only when levels of risk are high and unavoidable, will all help to reduce the costs of biosafety systems in Africa.

Fast-Track Options

There has always been the opinion that as regulators develop experience with GMOs, there will come a time when the regulatory oversight can be decreased on familiar products with a history of safe use. Yet, 20 years after the first approval of a genetically modified crop, there is little sign of the regulatory oversight decreasing in response to familiarity. In the mid- 2000s, the United States Environmental Protection Agency looked like it might be moving towards relaxation of regulatory oversight on virus-protected plants modified with coat protein genes, but this has stalled. Certainly, African regulatory authorities are well-advised to use existing food and feed safety data on GMOs that have prior approval in other countries. There is little value in collecting a new set of food and feed safety data in each country. Instead, where consumption patterns differ, or local communities have specific vulnerabilities, a sub-set of data could be collected to address these specific concerns.

Regional Biosafety Processes

Much of the review process for GMOs is repeated in each country in which the GMO will be tested or used. This is a considerable duplication of effort, and does little to increase the safety of GMOs. Efforts to establish workable regional biosafety review processes in Africa [e.g., Economic Community Of West African States (ECOWAS) and Common Market for Eastern and Southern Africa (COMESA)] should be encouraged, especially where they will reduce duplication, allow the sharing of technical resources, and enable countries without extensive biosafety infrastructure to gain access to science-based safety recommendations that will inform national decision-making on specific GMOs. The regional process would also function to address trade issues (Kimani and Gruere 2010), and ensure that there is minimal disruption to regional trade when GMOs are approved for use somewhere in the region.

Conclusion

If African farmers and consumers are to benefit from the products of modern biotechnology and these tools are to be used to drive economic development, food security, and poverty alleviation in Africa, then countries in the region need to ensure that their national biosafety frameworks for the approval of GMOs are functioning effectively. Careful consideration of the pitfalls of unworkable biosafety regulations are a powerful learning tool for countries that wish to move ahead with approvals for local testing and adaptation of safe and appropriate GMOs. Ensuring that the elements of workable regulatory systems are present and functioning nationally will greatly facilitate the move towards equitable access to improved technology and benefit-sharing across the continent.

References

African Union (2001) African model law on safety in biotechnology. http://www.africa-union.org/root/au/auc/departments/hrst/biosafety/AU_Biosafety_2b.htm. Accessed 1 Nov 2012

African Union (2011) Revised African model law on biosafety. http://www.au.int/en/dp/hrst/sites/default/files/2011-FinalDraftAMLS-en.pdf. Accessed 13 Nov 2012

Abramson SH, Van der Meer L (2002) Model biosafety Act. In: Biosafety regulation source book (2006) Arent Fox, Washington, DC. http://www.arentfox.com/modelbiosafetyact.pdf Accessed 1 Nov 2012

ICGEB (2011) Guidance for a fit-for-purpose regulatory framework for GMOs. Consensus document. International Centre for Genetic Experimentation and Biotechnology, Cape Town
James C (2012) Global status of biotech/GM crops: 2011, vol 43, ISAAA briefs. International Service for the Acquisition of Agri-Biotech Applications, Ithaca
Kimani V, Gruere G (2010) Implications of import regulations and information requirements under the Cartagena Protocol on biosafety for GM commodities in Kenya. AgBioForum 13 (3):222–241
McLean MA, Frederick RJ, Traynor PL, Cohen JI, Komen J (2002) A conceptual framework for implementing biosafety: linking policy, capacity and regulation, vol 47, ISNAR briefing paper. International Service for National Agricultural Research, The Hague
Morris EJ, Koch M (2002) Biosafety of genetically modified crops—an African perspective. AgBiotechNet 4:ABN 102
Raybould A (2006) Problem formulation and hypothesis testing for environmental risk assessments of genetically modified crops. Environ Biosafety Res 5:119–125
Traynor P, Frederick R, Koch M (2002) Biosafety in agricultural biotechnology. A workbook for training in biosafety risk assessment. Agricultural Biotechnology Support Project (ABSP), Michigan

Chapter 12
Harmonization of Regional Biosafety and Regulatory Services to Remove Future Trade Barriers in the COMESA Region

Getachew Belay, Virginia Kimani, and David Wafula

Abstract The Common Market for Eastern and Southern Africa (COMESA) is the largest regional trade bloc on the African continent, consisting of 19 member States, with a population size of about 400 million people. Agricultural products are an important part of trade in the COMESA region. There are several products that are of potential importance, as they are available globally as GM crops are/or being developed in some COMESA countries and Africa in general. Putting in place functional biosafety regulatory regimes is a prerequisite for the introduction and trans-boundary movements of GM crops. In most COMESA member States, the regulatory requirements for trade in GM-crop commodities are unclear. The COMESA Ministers of Agriculture have long realized the need for a regional approach, and launched the Regional Approach on Biotechnology and Biosafety policy in Eastern and Southern Africa (RABESA) Project in 2003. One of the major achievements of the RABESA project is the drafting in 2009 of the regional policy on cultivation, trade, and emergency food aid concerning GM crops. The process of policy formulation has undergone intensive consultations with key stakeholders in an inclusive, participatory, and interactive manner. The draft policy has gained support from the regulatory authorities and technical experts, and now awaits endorsement by the COMESA policy organs.

Disclaimer: The views and opinions expressed in this article are those of the authors, and do not necessarily reflect the official policy or position of the above institutions.

G. Belay (✉)
Alliance for Commodity Trade in Eastern and Southern Africa, COMESA Secretariat, Ben Bella Road, P.O. Box 30051 Lusaka, Zambia
e-mail: GBelay@comesa.int

V. Kimani
Pesticides and Agricultural Resource Centre, P.O Box 55191–00100, Nairobi, Kenya
e-mail: parcpest09@gmail.com

D. Wafula
Program for Biosafety Systems, P.O Box 1832–00100, Nairobi, Kenya
e-mail: wafuladavid@yahoo.com

F. Wambugu and D. Kamanga (eds.), *Biotechnology in Africa*, Science Policy Reports 7, DOI 10.1007/978-3-319-04001-1_12,

Keywords Biosafety • COMESA • Biotech/GM-crop trade • Regional policy harmonization

Abbreviations

ACTS	African Centre for Technology Studies
AGOA	African Growth and Opportunity Act
ASARECA	Association for Strengthening Agricultural Research in Eastern and Central Africa
AU	African Union
AU/NEPAD	African Union/New Partnership for Africa's Development
BCH	Biosafety Clearing-House
CAADP	Comprehensive Africa Agricultural Development Program
COMESA	Common Market for Eastern and Southern Africa
CPB	Cartagena Biosafety Protocol
EAC	East African Community
FFP	Feed, Food, or Processing
FTA	Free trade area
GM	Genetically modified
GMO	Genetically modified organism
IFPRI/PBS	International Food Policy Research Institute/ Program of Biosafety Systems
ISAAA	International Service for the Acquisition of Agri-biotech Applications
LMO	Living modified organism
LMO-FFP	Living modified organisms as food or feed, or for processing
NTB	Non-tariff barriers
PoE	Panel of Biotechnology and Biosafety Experts
RABESA	Regional Approach on Biotechnology and Biosafety policy in Eastern and Southern Africa
REC	Regional Economic Community
SADC	Southern African Development Community
SSA	Sub-Saharan Africa
USAID	United States Agency for International Development
WTO	World Trade Organisation

Introduction

The Common Market for Eastern and Southern Africa (COMESA) region (Fig. 12.1), the largest trading block in Africa, comprises 19 member states with a population size of ca. 400 million (COMESA 2012). Agriculture still remains the

Fig. 12.1 COMESA member states

most important sector of the economies of the Member States in terms of GDP contribution, employment, and foreign exchange earnings. However, the region is grappling with high food prices, as a result of ever-escalating prices of fuel and agricultural inputs and competition with bio-fuels, and global climate change adds to the overall decline.

The growth of the agriculture sector depends on productivity and access to markets. The Comprehensive Africa Agricultural Development Program (CAADP), which was launched under the AU/NEPAD, is being implemented in many African countries, including the COMESA member States, to manage the growth of the agricultural sector. The CAADP framework recognizes that growth of the agricultural sector, among other things, requires supportive public policies, favorable investment climate, national and regional market access, and use of appropriate technologies.

Seventeen years have passed since the first GM crop was commercialized. The cultivation of GM crops has now reached 170 million ha in 28 countries globally (James 2012). Many countries have also developed their regulatory frameworks based on the international regulatory regime, the Cartagena Protocol on Biosafety.

Among the COMESA Member States, Egypt was the first to achieve commercial release, with Bt maize, followed by Bt-cotton in Sudan in 2012. Several product-testing trials are being carried out in Kenya and Uganda, including disease/pest-resistant varieties of African crops; for instance, resistance to banana wilt disease, cassava mosaic virus, striga-weed in sorghum, and insects in cowpea.

Application of modern biotechnology is not the panacea to addressing the multi-faceted development challenges in African agriculture, but is a powerful option for consideration. The COMESA Member States are at different stages in biotechnology policy and capacity development. However, GMO issues have become increasingly relevant to the region, because biotechnology products are already in food chains, trade, and food-aid.

The utilization of GM crops in agricultural production and trade cannot be carried out without a suitable biosafety regulatory system. Both the regulatory requirements and biotech-product development are knowledge-intensive and resource-demanding. This was the main reason that the COMESA Ministers of Agriculture decided, in Uganda in 2001, to handle biotechnology through a regional approach. Subsequently, in 2003, the Regional Approach to Biotechnology and Biosafety in Eastern and Southern Africa (RABESA) initiative was launched, to implement the Ministerial decision.

One of the key tasks of the RABESA project has been the drafting of a COMESA-wide biotechnology and biosafety policy framework. This chapter describes the needs that led to a regional approach, the experiences gained in the policy formulation process, and the impact of the Project since inception.

Overall Trade Potential for COMESA

COMESA is a free trade area (FTA) that was launched in the year 2000. Member States that currently belong to the FTA trade on a duty-free and quota-free basis among themselves, provided that the goods meet the COMESA rules of origin.

The three Regional Economic Communities (RECs), COMESA, SADC, and EAC also have a Non-Tariff Barriers Monitoring Mechanism. In the region, Member States of the SADC, COMESA, and EAC have decided to adopt a harmonized approach with regard to the elimination of non-tariff barriers (NTBs). The three RECs have joined forces to implement a common NTB reporting, monitoring, and eliminating mechanism, which incorporates concrete timelines for the removal of NTBs in the region.

Operators can directly report and monitor the resolution of barriers encountered in the COMESA, EAC, and SADC regions online. This new system enhances transparency and makes it easy to follow up reported and identified NTBs. This web-based NTB system is accessible to all economic operators, public officials, academic researchers, and other interested parties.

In addition, the three RECs are in the process of negotiating a tripartite agreement for an FTA. The draft document has been prepared and is currently under discussion. The 26 tripartite countries include Angola, Botswana, Burundi, Comoros, Democratic Republic of Congo, Djibouti, Egypt, Eritrea, Ethiopia, Kenya, Lesotho, Libya, Madagascar, Malawi, Mauritius, Mozambique, Namibia, Rwanda, Seychelles, South Africa, Sudan, Swaziland, Tanzania, Uganda, Zambia, and Zimbabwe.

The trade potential of the three RECs is understood to be substantial. In 2008, estimates indicated that exports among the 26 tripartite countries were USD 27 billion and imports were USD 32 billion. The combined GDP for the same year was estimated at $624 billion (Mwapachu 2009).

Trends in Agricultural Commodity Trading in the COMESA Region

Trade in agricultural products in the region has advantages (Karim Abdel and Ismail 2007). First, trade can contribute to stabilizing supply when national fluctuations in production are greater than the fluctuations in the region. Thus, free intra-regional trade among the COMESA countries could be an efficient substitute for national stockpiling, and might be used to even out fluctuations in national production. It has been suggested that worldwide free trade in grains would drastically reduce the need for holding carryover stocks, because fluctuations in world cereal production are minimal compared with fluctuations in national production [Johnson (1981), quoted in Abdel Karim and Ismail (2007)]. The same may hold true if variability in production in individual member countries is greater than variability in production for the COMESA region as whole.

Second, trade in agricultural products may partly substitute for working stocks if the harvesting calendar differs somewhat among trading partners. Third, trade may allow countries to specialize in production in accordance with comparative advantage. Thus, trade would help to increase national income and improve food security.

The overall value of total COMESA product imports in 2011 was USD 132 billion. Exports stood at USD 96 billion. The value of imported services was estimated at USD 21 billion, while exports were valued at USD 28 billon (Figs. 12.2 and 12.3). The share of goods imported by COMESA in 2011 accounted for 0.7 % of the world imports.

There are several products that are of potential importance, as they are available globally as GM crops. These include maize, soybean, cotton, and canola. Cassava, banana, and sorghum are being developed in some countries in the COMESA, and Africa in general.

Total maize (grain and seed) bilateral trade in the COMESA region was valued at USD 218 million, while imports from the world were valued at USD 2.5 billion (Fig. 12.4). There is consistent demand for maize, which is currently met by the rest of the world. In 2011, the main exporting countries within COMESA were Malawi, Zambia, and Uganda. The major importers were Egypt and Kenya.

There is minimal trade in canola or rapeseed. Mauritius is the most significant importer of canola, and in 2011 it imported 81 % of the total valued at USD 0.29 million. Imports of vegetable oil, particularly palm oil, into the COMESA region are highly significant. This may be replaced by canola due to the global health shift from saturated to unsaturated oils.

Cotton seed trade has been significant in the region. Seeds are traded whole, broken, as oil fractions, and as seedcake. These are mainly used for the manufacture of animal feeds.

Sorghum is an important staple food in the COMESA region. While more than ten countries export and import grain sorghum, imports exceed exports. Sorghum has generally been regarded as a traditional crop, and most of it is consumed in the countries where it is grown (FAO 1988). Large imports of wheat and rice in some

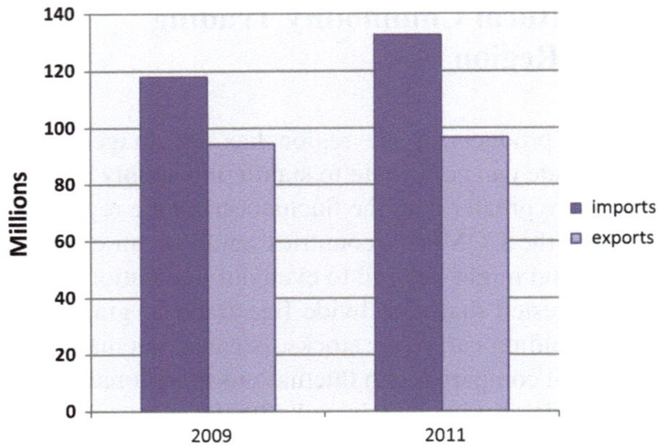

Fig. 12.2 COMESA value of trade in 2009 and 2011 (USD) (Source: Compiled by author from UNCOMTRADE 2012)

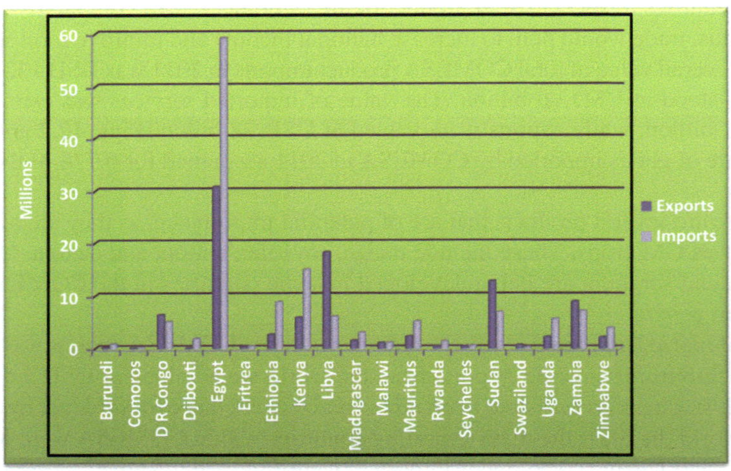

Fig. 12.3 COMESA countries trade values in '000 USD 2011 (Source: Compiled by author from UNCOMTRADE 2012)

countries, as well as urbanization, has had considerable negative impact on the production and utilization of sorghum. However, efforts are on to increase production of the crop. In this regard, the goal of the Africa Biofortified Sorghum (ABS) project is to develop a transgenic sorghum that contains increased levels of essential nutrients, especially lysine, Vitamin A, iron, and zinc. The nutrition-enhanced sorghum will be used for introgression of the nutritional traits into high-yielding African and farmer-preferred varieties (ABS 2012).

Currently, among the COMESA countries, Sudan is the most important importer of sorghum, followed by Ethiopia. The countries that mainly export sorghum are

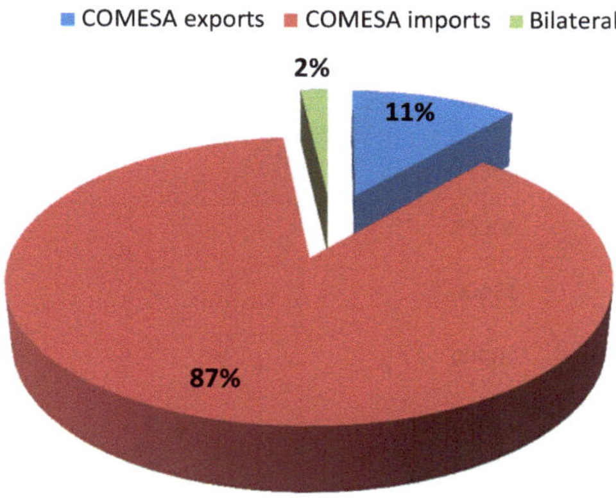

Fig. 12.4 Maize imports from the world, exports to the world, and bilateral trade in COMESA (Source: Compiled by the author from UNCOMTRADE 2012)

Argentina, India, Italy, Russia, and USA. However, in 2011, Sudan and USA were the predominant sources of imported sorghum into the COMESA region.

There is a significant trade in sorghum among the COMESA countries and in Africa as a whole (Fig. 12.5). However, this depends on the availability of surplus, which may be achieved with the development and adoption of biofortified sorghum. There is also significant potential for the use of diversified sorghum, such as for feed and beer making. The proportion of agricultural imports out of overall imports is variable for different countries.

Agricultural imports into Egypt are dominated by food, especially cereals (constituting 8 % of the total imports), oil seeds (1 %), sugar (1 %) beverages, fruit, and nuts (all <1 %). The specific cereals imported in decreasing importance are wheat, maize, rice, and grain sorghum. Others are oats, barley, and rye. Egypt also imports 1 % of its cotton demand.

The dominant agricultural imports into Egypt constitute biotech crops, including maize, oil crops, cotton, sugar, and sorghum. The current sources of these commodities are countries that have adopted GM crops, including Argentina and USA. Maize is one such example, and the import trend from the top nine countries is shown in Fig. 12.5. Almost all the maize imports are sourced from outside the COMESA block (Fig. 12.6). The demand for maize in Egypt far outstrips supply from Africa, even with South Africa supplying about 2 % of the commodity.

Oil seeds and related commodities also have potential for trade within COMESA. Much of the soybean oil and meals are imported from GM producers, chiefly Argentina, Brazil, Canada, and USA. Soya meal for animal feed and aquaculture is preferred due to higher protein content and processing in the US. As is the case with cotton, Sudan is also a significant supplier of soybean, but the proportion is quite low compared with the total import volumes. For example, Sudan exports are only about 4 % of the total imported value from Argentina. There is obviously significant potential for increased soybean trade between Sudan and Egypt. South Africa is a minor supplier, and exported soybean worth USD 388,000 USD in 2011 (UN COMTRADE 2012). This material is most likely GM soybean.

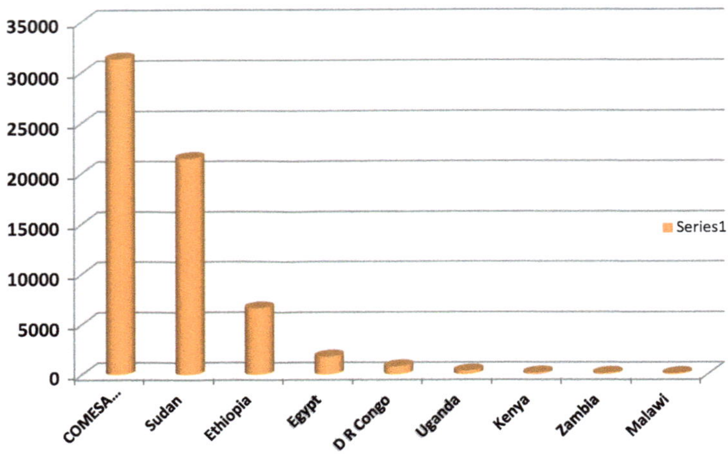

Fig. 12.5 Importers of sorghum in the COMESA region. '000 USD (2011) (Source: Compiled from UNCOMTRADE 2012)

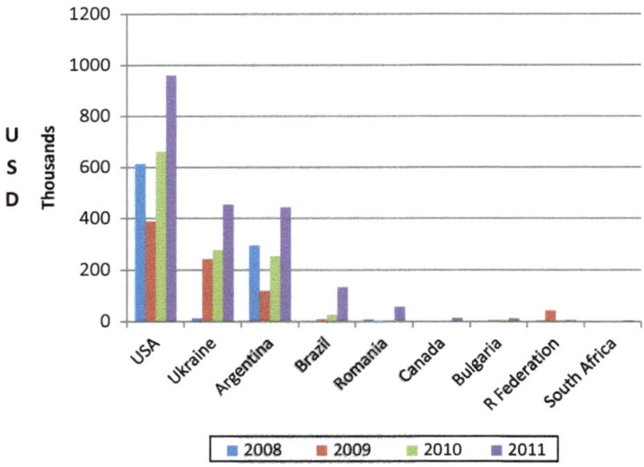

Fig. 12.6 Value of maize imports into Egypt (2008–2011)

Food-based exports from Egypt include fruits (3 %), edible roots and tubers (3 %), cotton (2 %), and several other agricultural products. Exports of cotton to the COMESA region have been declining, as shown in Table 12.1.

Egyptian cotton production has been facing a crisis, as farmers could not get good prices for their crop (Mansour 2011). The government has not been willing to subsidise, due to WTO rules. This has led to an overall decline in production, the local demand being met more via imports. Bt cotton is still under development, and possibly offers an opportunity for farmers to cut costs

Table 12.1 Exports of cotton into COMESA from Egypt compared with imports from the rest of the world-value in '000 USD

	2009	2010	2011
Egypt exports to COMESA	981	1,087	450
Overall COMESA imports	1,019,915	–	1,247,446

and offer a cheaper commodity to both Egyptian processors and countries within COMESA.

The share of agricultural imports into Kenya is significantly food-based, including categories covering food and live animals, beverages and tobacco, animal and vegetable oils and fats, oil seeds, oil nuts, and oil kernels. In 2010, these categories contributed 12 % of the imports. In the food category, cereals such as maize, wheat, and rice contributed the most, accounting for 4 % of the imports, with a value of about USD 600 million.

Imports of maize in 2011 mainly originated from Malawi and Zambia. This was a shift from 2008 and 2009, when imports were mainly from South Africa. Conflicts attributable to unclear legislation and its application resulted in rejection of maize consignments from South Africa. As a result, multiple small consignments were imported from Malawi, Zambia, Tanzania, and Uganda in the following 2 years (Fig. 12.7). Malawi and Uganda have non-GM maize, and were found suitable alternative sources to South Africa, which has commercialized GM maize, cotton, and soybean.

Kenya is also a significant importer of soybean products. Much of the soy is imported as corn–soy blend, while some fraction is used in animal feeds. Soy meal was imported mainly from South Africa (Fig. 12.7). These fractions are likely to be genetically modified.. There is a consistent demand for soy, though the volumes are not very big. In an earlier study, Kimani and Guillaume (2010) concluded that as the global importance of soybean continues to increase, it is expected that Kenya's demand will also increase proportionately, laying the foundation for vibrant trade within COMESA countries.

Unprocessed soy, however, is mostly imported from China, Uganda, and USA. Other commodities with similar potential are cotton, especially with the opportunities offered by AGOA and canola, a likely replacement for palm oil.

Ethiopia is the third largest COMESA importer from the world with a total value of USD 8 million. The main sources of their goods in 2011 included China, India, Saudi Arabia, the United Arab Emirates, and USA. The main agricultural goods imported were cereals, fats, and oils. Among the cereals, the most significant Ethiopian imports are wheat and grain sorghum. In 2011, wheat was imported from the Russian Federation and USA. Grain sorghum was exclusively imported from Sudan. Smaller volumes of maize came from South Africa and USA (UN COMTRADE 2012).

Fig. 12.7 Import values of maize into Kenya, 2007–2011 ('000 USD)

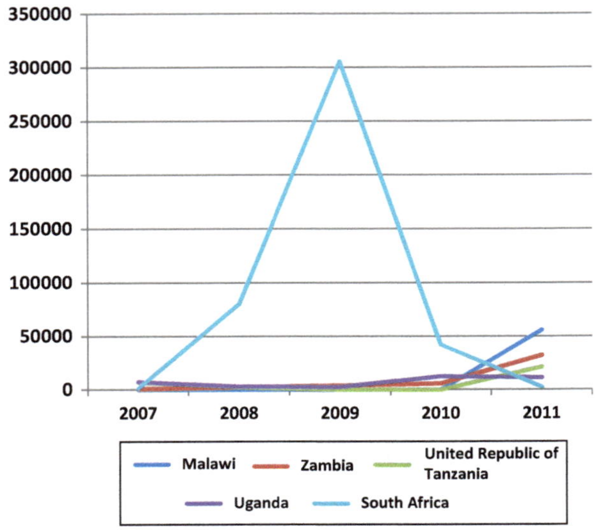

Barriers to Regional Trade in GM Commodities

Regulatory barriers: Most developing countries, particularly in Africa, have ratified the Cartagena Protocol, and consider it the reference point in the setting of their own biosafety regulations (Gruere 2006; Jaffe 2006). In the Protocol, there are specific rules related to the approval and documentation of imported Living Modified Organisms (LMOs) intended for direct uses as Feed, Food, or Processing [noted LMO **as Food or Feed, or for Processing** (FFPs)], essentially unprocessed GM commodities. Under Article 18.2.a., parties to the Cartagena Biosafety Protocol (CPB) "should request information" from exporters regarding the presence and the identification of LMO-FFPs in any shipment before importation (Advanced Informed Agreement).

Under this rule, shipments containing LMO-FFPs identified "through means such as identity preservation systems" must show that the shipment "does contain" LMO-FFPs, and provide a list of GM events present in the shipment. Shipments of LMO-FFPs that are not well-identified will only have to label their shipment as "may contain LMO-FFPs," and information on the complete list of GM events commercialized in the exporting country would be available to the importers via the Biosafety Clearing-House (BCH) or if not, on the importers' request (Redick 2007).

In most COMESA member States the focus on biosafety has been on introductions and release of GM crops through cultivation. Therefore, the regulatory requirements for trade in GMO commodities are fairly unclear. Further, signatories to the Protocol have been reviewing the initial provisions regarding trade in LMOs-FFPs, and the final agreement on imports procedure is still unclear. This has resulted in a lack of clear process in trade licensing, including in the identity of the responsible authority.

Policy versus *legislation conflict*: The opinions on the importance and safety of GM crops in developing countries have continued to be varied, and at times negative. Individual policy makers are often also undecided on the safety of the crops at a personal level. As a result, there is occasional conflict between policy as defined by the political class, and the legislation as implemented by the regulatory authorities and overall government position.

Standards and custom requirements: Globally, according to the Harmonised Codes set by the World Customs Organization, GM crops are not listed any differently than conventional products. Due to the general concerns regarding GMOs, the commodities undergo double clearance where the policy and regulatory procedures have not been clarified to the customs authorities.

There are often multiple regulatory clauses considered relevant to biotech crops and GMOs, resulting in differing requirements by various actors. An important example is the definition of quality. In most cases, quality attributes are set by the national standards organization. Frequently, the standards making process does not involve the biosafety regulators. In addition, the standards body also sets quality conditions to be met for pre-import inspection. The biosafety authority may therefore decline imports already approved at pre-shipment.

COMESA's Regional Approach on Biotechnology and Biosafety

The importance of regional cooperation in harnessing modern biotechnology safely and responsibly, and handling of other GMO-related issues, is evident from the experience of other global regional blocs. As countries within COMESA implement policies on biosafety and biotechnology at the national level, it becomes more and more important to adopt harmonized biosafety policies at the regional level. Products that have been tested and approved in one Member State would not have to undergo further testing before adoption; this will promote adoption of the products, result in increased trade, and facilitate access to emergency food aid. Of particular importance are food commodities such as maize and soy bean, as well as animal feed raw materials such as cotton seed.

The need to pursue a regional approach was mooted by the COMESA Ministers of Agriculture at a meeting held in 2001 in Kampala, Uganda. The COMESA Secretariat took the necessary steps to conceptualize the resolutions of the Ministers, and consequently designed the RABESA project aimed at responding to the critical concerns identified. The Ministers endorsed implementation of the project in 2003, and this paved way for commencement in 2004.

The COMESA regional harmonization approach takes its legitimacy from the COMESA Treaty. The provision of Article 129 of the Treaty stipulates full cooperation in agricultural development, science and technology domains, to increase agricultural production and attain regional food security. Further, Article 130(a) of the COMESA Treaty stipulates that Member States undertake to co-operate in specific fields of agriculture, including the harmonization of agricultural policies of the Member States with a view to having a common agricultural policy.

Achievements of the RABESA Project

The RABESA project has been implemented in two distinct but interrelated phases focused on the ultimate goal of bringing together COMESA Member States to cooperate in handling biosafety issues at the regional level. In Phase I (2004–2007), three policy studies were conducted to generate evidence required to support realization of the project objectives:

(i) potential farm-income gains from the adoption of GM crops,
(ii) the magnitude of commercial export risks associated with GM crops; and
(iii) the delivery of emergency food aid with GM content in the COMESA region

The study on farm-income gains projected that COMESA Member States could harness substantial benefits from the adoption of GM insect-resistant varieties of cotton and maize (Paarlberg et al. 2006a). The most innovative and ground-breaking research focused on trade-related implications of adopting GMOs in the COMESA region. RABESA was the first initiative in Africa to demystify the magnitude of export risks associated with GMOs in a concrete way. The key conclusion was that inter-regional export losses associated with the adoption of GM crops in the COMESA region were negligible. Although COMESA countries depend heavily on the export of agricultural products to earn foreign exchange, the major exports being coffee, tea, sugar, horticulture, banana, and pyrethrum, none has been commercialized anywhere in GM form, meaning there is currently little or no GMO-associated risk to agricultural export incomes (Paarlberg et al. 2006b).

The food aid policy study revealed that sub-Saharan Africa (SSA) is the largest recipient of emergency food aid globally, and COMESA countries receive 85 % of all emergency food aid to SSA. About 50 % of the food aid arrives as in-kind donations from countries that are leading producers of GM crops, including USA and Canada (Paarlberg et al. 2006c).

The Policy research findings were disseminated and discussed at national and regional-level consultative meetings and workshops, where a consensus was built on priority areas of harmonization; commercial planting of GMOs, trade in GM products, and delivery of emergency food aid with GM content.

The 4th meeting of COMESA Ministers of Agriculture, held in 2007, endorsed the development of regional biosafety policies and guidelines focusing on the three identified areas of harmonization, and passed a decision to form an interim Panel of Biotechnology and Biosafety Experts (PoE) to serve as an advisory body of COMESA in guiding the harmonization process. These Ministerial decisions paved the way for commencement of RABESA phase II.

In March 2009, the COMESA Secretariat initiated the drafting of COMESA Regional Biosafety Policies and Guidelines. A Biosafety Roadmap and a Communication Strategy were also drafted. The rationale behind development of a Biosafety Roadmap was explained by the fact that existence of functional biosafety systems at the national level is a key requirement for countries to align to and optimize the benefits of a regional biosafety framework. The Roadmap is expected to encourage and guide the implementation of national biosafety frameworks by more countries in the COMESA region, and ensure that countries are striving to achieve common goals. The COMESA communication and advocacy strategy seeks to support and create awareness of the benefits associated with regional harmonization.

The policies and guidelines, the Roadmap and Communication Strategy drafts were subjected to several rounds of technical review and stakeholder consultations. Subsequently, COMESA Ministerial meetings held in Zambia in 2010 and in Swaziland in 2011 provided further guidance in terms of the approach that should be taken to promote consensus and effective stakeholder participation at the national level in all the Member States. In implementing the decision of the Ministers, the COMESA Secretariat conducted 18 national workshops between September 2010 and February 2012. Libya was the only exception because of heightened civil unrest at the time.

The policies and guidelines were revised systematically to reflect comments and inputs from key stakeholders, regulatory authorities, and technical experts, and validated in a regional workshop in 2012, in Lusaka, Zambia. The pivotal provisions included in the draft policy are: (a) collective recognition of both the benefits and potential risks associated with GMOs on a case-by-case approach, (b) a regional-level and science-based biosafety risk assessment mechanism, coupled with national level decision-making, and (c) capacity building. Briefly, the COMESA regional harmonization is all about sharing information, resources, and expertise, avoiding redundancy and reducing cost of biosafety regulations. The process of policy formulation has undergone intensive consultations with key stakeholders, including civil society organizations, in an inclusive, participatory, and interactive manner. The draft policy has gained support from the regulatory authorities and technical experts, and now awaits endorsement by COMESA policy organs[1].

Impacts and Outcomes of the RABESA Project

The RABESA Project has been recognized by the African Union (AU) as one of the outstanding models of regional harmonization in biosafety on the continent. While several studies on the potential farm-level impacts associated with adoption of GM crops have been conducted in Africa, the contribution of RABESA in breaking new ground and unravelling trade-related impacts and the emergency food aid dimension of GMOs was innovative and valuable.

[1] The COMESA Policy on Biotechnology and Biosafety has been adopted by the 32nd Meeting of the Council of Ministers, 23–24 February 2014, Kinshasa, Democratic Republic of Congo.

The creation of a biotechnology and biosafety unit within the COMESA Secretariat is a major spillover of the Project. The unit has conspicuously emerged as the COMESA regional focal point on biotechnology and biosafety issues. Apart from the RABESA Project, COMESA has taken a leading role in supporting other regional processes, such as strengthening the capacity of African delegates to understand and negotiate effectively the issues surrounding the implementation of the Cartagena Protocol on Biosafety.

Lessons Learnt from Implementation of RABESA Project

Important lessons have been learnt during the implementation of the RABESA Project. Key among these are:

- Issues of regional harmonization should be handled in a consultative, participatory, and inclusive manner. This is because, given the controversies, the process of policy formulation is no less important than the policy framework itself. Deliberations cutting across the entire life cycle of the RABESA Project have taken place in 24 national and four regional workshops.
- Regional harmonization of biosafety policies is both a technical and political process that requires strong political will and commitment at various levels within Member States. The progress made and political buy-in realized so far is attributed to the fact that the RABESA Project has been one of the key and recurrent agenda items in various COMESA policy organ meetings.
- National sovereignty is a fundamental and sensitive issue. The convergence and divergence between national and regional frameworks has to be clearly spelt out, and pertinent concerns handled carefully to dispel fears that the regional process may infringe on or override national interests and decision-making powers.
- Awareness and outreach efforts need to be stepped up in order for countries to appreciate the benefits of a harmonized approach in biosafety decision-making. This emphasises the need for a focused and demand-driven communication and advocacy strategy.

Concluding Remarks

COMESA's mission statement is to provide excellent technical services to its Member States, in order to facilitate the region's sustained development through economic integration. Since the launch of the COMESA Free Trade area in the year 2000, there has been a steady increase in formal and informal intra-COMESA trade in agricultural products. In order to realise increased gains in intra-COMESA trade, harmonization of biosafety policies and their rational implementation will play a significant role in social and economic development within the block.

Although the level of opposition to GM products in the region is still considerably high, many of the arguments and counter-arguments give more attention to field-cultivation of GM crops rather than trade. The types of crops for which GM products are available (maize and soybean), and those in the pipeline (mainly banana, cassava, sorghum, cowpea) are relevant and important in the African continent. With such porous borders among Member States, it is highly likely that the arrival of GM crops in one country will eventually make its way to another. In the long run, it is possible that, unless harmonized, the GM factor may rise to the level of trade disruption among COMESA countries themselves. Therefore, focusing on uncompromised region-wide biosafety risk-assessment instruments, in a complementary fashion with national-level mechanism, would be a realistic way forward. The idea is simple: without infringing national-level decisions; if one country does the risk-assessment properly, it should not necessarily be repeated in all the Member States.

Having the policy framework is just the beginning. More challenges lie ahead when it comes to turning policy into practice. Regulatory capacity-building needs of Member States should be given serious considerations and concerted efforts. Therefore, implementation of a regional initiative of RABESA's magnitude calls for strong and sustained partnerships. Right from the onset, COMESA engaged strategic partners with varied strengths and competencies to support implementation and harness complementarity. The diversity and the status of the partnerships also enhance the profile and credibility of the Project and also the process. High levels of commitment, consistency, and patience from the partners are indispensable for protracted regional harmonization processes such as the RABESA Project.

Acknowledgements The COMESA Biotechnology and Biosafety Program has been financially supported by the United States Agency for International Development (USAID). The partners that have supported COMESA in the implementation of the RABESA Project since inception include the Association for Strengthening Agricultural Research in Eastern and Central Africa (ASARECA), the Program for Biosafety Systems [International Food Policy Research Institute (IFPRI)/Program of Biosafety Systems PBS), the International Service for the Acquisition of Agri-biotech Applications (ISAAA *AfriCenter*), and the African Centre for Technology Studies (ACTS).

References

ABS (2012) Africa biofortified sorghum. ABS project: technology development. http://biosorghum.org/home.php. Accessed 8 Oct 2012

COMESA (2012) About COMESA. http://about.comesa.int/. Accessed 8 Oct 2012

FAO (1988) Regional trends in production and utilization of sorghum and millets. http://www.fao.org/docrep/T0818E/T0818E00.htm#Contents. Accessed 8 Oct 2012

Gruere G (2006) An analysis of trade related regulations of genetically modified food and their effects on developing countries. EPT discussion paper 147. International Food Policy Research Institute, Washington, DC

Jaffe G (2006) A comparative analysis of the national biosafety regulatory systems in East Africa. EPTD discussion paper 146. International Food Policy Research Institute, Washington, DC

James C (2012) Global status of commercialized biotech/GM crops: 2012. ISAAA, Ithaca. ISAAA brief no. 44

Karim IEEA, Ismail IS (2007) Potential for agricultural trade in COMESA region: a comparative study of Sudan, Egypt and Kenya. Afr J Agric Res 2(10):481–487

Kimani V, Gruère G (2010) Implications of import regulations and information requirements under the Cartagena Protocol on biosafety for GM commodities in Kenya. AgBioForum 13 (3):222–241

Mansour S (2011) Foreign Agricultural Service. Global Agricultural Information Network, 2011. Cotton Update. http://gain.fas.usda.gov/Recent%20GAIN%20Publications/Cotton%20update_Cairo_Egypt_10-20-2011.pdf. Accessed 15 Aug 2012

Mwapachu JV (2009) State of the tripartite free trade area: progress (from the report of the chairperson of the Tripartite Task Force — Amb. Juma V. Mwapachu — Secretary General, EA). http://www.comesa-eac-sadctripartite.org/intervention/focal_areas/tripartite_fta. Accessed 13 Sept 2012

Paarlberg R, Wafula D, Minde D, Wakhungu J (2006a) Projected farm income gains in the COMESA/ASARECA region from commercialization of Bt maize. ACTS Press, Nairobi

Paarlberg C, Wafula D, Minde D, Wakhungu J (2006b) Commercial export risks of genetically modified (GM) crops in the COMESA/ASARECA region. ACTS Press, Nairobi

Paarlberg R, Wafula D, Minde D, Wakhungu J (2006c) Food aid import policies in the COMESA/ ASARECA region: the costs and benefits of current policy options. ACTS Press, Nairobi

Redick TP (2007) The Cartagena protocol on biosafety: precautionary priority in biotech crop approvals and containment of commodities shipments. Colo J Int Environ Law Policy 18:51–116

UN COMTRADE (2012) Trade statistics for international business development. http://www.trademap.org/SelectionMenu.aspx. Accessed various dates, July–Oct 2012

Part VI
Communication and Community Engagement

Chapter 13
Why Communication and Issues Management (CIMS) Must Occupy a Central Role in GM Projects: Case Study of the Africa Biofortified Sorghum (ABS) Project

G. Daniel Kamanga, M. Florence Wambugu, Silas Obukosia, Rose Gidado, and Iro Suleiman

Abstract Since the commercialization of the first transgenic crop nearly 2 decades ago, the areas planted to biotech crops have steadily increased through the years; this has not been without controversy. To deal with this, most biotech projects have had to develop communication strategies all the way from research and technology development to commercialization. For private sector biotech companies, communication is part and parcel of normal business, but even they, like many public and non-governmental organizations (NGOs), have been forced to place greater emphasis on communication.

This paper explores the tension between technology development and communication. It argues that communication should be involved "from lab-to-fork." The experience of the Africa Biofortified Sorghum (ABS) Project is used as a case study of the importance of communication, in light of the arguments that there is nothing to communicate if there is no product. It underlines the need for emphasis of the communication function at different phases of a biotech project.

For the ABS Project, continued success has hinged on the project's ability to balance technology development and communication. This paper looks at the

G.D. Kamanga (✉)
Africa Harvest Biotech Foundation International (AHBFI), 34 Forbes Road, Blairgowrie 2194, South Africa
e-mail: dkamanga@africaharvest.org

M.F. Wambugu
Africa Harvest Biotech Foundation International (AHBFI), Nairobi, Kenya

S. Obukosia
Africa Harvest Biotech Foundation International (AHBFI), P.O. Box 642, Village Market 00621, Nairobi, Kenya

R. Gidado
National Biotechnology Development Agency (NABDA), Abuja, Nigeria

I. Suleiman
Institute of Agriculture Research (IAR), Zaria, Nigeria

F. Wambugu and D. Kamanga (eds.), *Biotechnology in Africa*, Science Policy Reports 7, DOI 10.1007/978-3-319-04001-1_13,

crafting of the ABS communication strategy, analyses its implementation, and proposes a way forward, not just for the ABS Project, but for similar biotech projects in Africa.

The paper concludes by proposing two ideas for the future: first, that private and public-sector biotech projects in Africa should consider forming an African Biotech Coalition (ABC) to create a powerful force to advance their cause, and second, that to build on the current momentum of public acceptance, every effort should be made to demonstrate the benefits of the technology.

Keywords Biotech communication • Communication • Communication strategy • Communication roadmap • Communication and Issues Management • Issues Management • Africa biotech communication • African biotech projects • ABS project • Anti-GM

Abbreviations

ABC	African Biotech Coalition
ABS	Africa Biofortified Sorghum
ACB	African Centre for Biosafety
AHBFI	Africa Harvest Biotech Foundation International
BMGF	Bill and Melinda Gates Foundation
CEO	Chief Executive Officer
CFT	Confined field trials
CIMS	Communication and Issues Management Strategy
CSIR	Council for Scientific and Industrial Research
DMU	Decision-making unit
FoEN	Friends of the Earth Nigeria and Environmental Rights Action
GC	Grand Challenge
GCGH	Grand Challenges in Global Health
GE	Genetic engineering
GHT	Greenhouse trials
GM	Genetic modification
GMO	Genetically modified organism
IAR	Institute of Agriculture Research (Nigeria)
KARI	Kenya Agricultural Research Institute
MNC	Multinational company
NGOs	Non-governmental organizations
PCT	Project Communication Team
PI	Principal Investigator
PSC	Project Steering Committee
TLMG	Team Leaders Management Groups

Introduction

Agricultural transgenic technologies—popularly referred to as "genetic modification (GM)", "genetically modified organisms (GMOs)" or "genetic engineering (GE)"—have generated a lot of controversy in Africa. African scientists and policymakers are now better informed. Yet, it is clear that increased communication has had both negative and positive effects. Africa has had its share of GM-related controversies.

While the controversies continue, Africa remains the only continent where per capita food production is decreasing. Hunger and malnutrition affect one in three Africans (Shannon 2010). However, the potential to increase agricultural productivity and alleviate poverty has caused South Africa to view the GM technology as "an integral part of SA's food security policy" (De Villiers 2013).

African countries are considering the GM technology because of high input costs of fertiliser, pesticides, and insecticides (FAO 2012). South Africa, Egypt, Sudan, and Burkina Faso are currently the only four countries that have commercial GM crops on the continent (Allen 2013). Kenya and Ghana have laws that allow commercialization, but no crops have been commercialized yet. Despite the apparent increase in biotech acceptance, sustained opposition to GM crops in Africa—mostly organised by international NGOs—shows no sign of slackening (Namibia Economist 2011). Much of the opposition is based on emotive arguments and thrives on the ignorance of most target audiences about biotechnology and the life sciences in general.

It was into this environment of fear and distrust towards biotechnology—and the GM technology specifically—that the Nutritionally Enhanced Sorghum for the Arid and Semi Arid Tropical Areas of Africa (otherwise called the Africa Biofortified Sorghum (ABS) Project) was launched into in 2005. It was one of a bouquet of projects, the result of Bill Gates' vision to find solutions to diseases that disproportionately affect the developing world. His solution was the Grand Challenges in Global Health (GCGH). Some 14 grand challenges (now refered to as GCs) were identified.

Grand Challenge 9, or GC#9, focused on agriculture, more specifically, "creating a full range of optimal bioavailable nutrients in a single staple plant species". In this GC, the Foundation was acknowledging that the malnutrition challenge is a major global health problem. Four projects were selected and funded with the sole focus of creating nutrient-rich staple crops. The ABS Project was one of the four projects. Africa Harvest provided leadership in putting together an African-led consortium, bringing together several African scientists and leading African research institutions to form the ABS Project (www.biosorghum.org).

The communication strategy was viewed as essential to the smooth running of the project. Previous experiences showed that successful implementation of similar projects was hampered by perceived scientific issues around the environment, food safety, and societal concerns revolving around public acceptance. One of the ABS Project partners, the Council for Scientific and Industrial Research (CSIR) had just been accused by anti-GM activists of using South Africa as a "guinea pig" for producing and testing "mutant AIDS drugs" (Gedye 2004). The fact that the attacks started even before the ABS project began underlines the need for a public acceptance strategy at the beginning, and throughout the life, of such projects.

The strategy-building process required studying issues related to public acceptance of similar projects. The introduction of GM maize in Mexico, a centre of diversity for the crop (Zietz and Seals 2006), and the Golden Rice Project (GRP) http://www.goldenrice.org provided the greatest inspiration to developing the ABS communication strategy. The twin-objective of the ABS Project's communication strategy was to pave the way for research and development while clearing the acceptance path for the final product. The design of the communication strategy was part and parced of the project concept submitted to the BMGF. Communication milestones and activities were designed to underpin the discovery, research and development, regulatory, and commercialization processes. The strategy relied on building alliances and capacity of critical stakeholders such as government officials, public institution scientists, and communities in target markets. In addition, the project undertook communication research to better understand and deal with key concerns, perceived or real.

The Communication and Issues Management Strategy (CIMS) recognized that, although the GC#9 projects were framed as "nutrition projects", anti-GM activists would seek to re-frame them. For example, Friends of the Earth Nigeria and Environmental Rights Action (FoEN)—referring to one of the GC#9 projects—"called for immediate end to GM cassava testing" (www.allafrica.com 2009). In Uganda, another GC#9 project, was said to "generate a wide portfolio of concerns, as the technology of genetic engineering is still in its early stages of development in Uganda." (Kikulwe et al. 2010)

Coming into a polarized debate, the ABS Project recognized that if it did not communicate with various target audiences, the anti-GM groups would exploit information gaps, modify, distort and/or manipulate information to support their ideological views and to win the support of critical role players and beneficiaries. The emerging strategy included a comprehensive plan to reach out to various stakeholders, who included grassroot communities, NGOs, scientists, policy makers, politicians, and the media. The outreach had three goals: provide information on the project R&D, address areas of concern, and encourage dialogue and consensus building.

The ABS Project started as a Consortium of nine institutions led by Africa Harvest. DuPont-Pioneer donated the initial technology, or ABS#1, which consisted of a sorghum product with 50 % higher lysine content. This was valued at US$4.8 million, representing the development of the genes and characterization of their function. The project built on this initial work and created ABS#2, whose target was to develop a highly fortified product with improved essential amino acid composition, protein/starch digestibility, iron and zinc availability, and elevated levels of select vitamins, including Vitamin E.

Organizationally, one of the challenges was to build communication and issues management capacity within member institutions so that the consortium members would operate a unified communication strategy and "speak in one voice" on key issues. Africa Harvest was the Consortium Leader and provided the Co-Principal Investigator (PI), Du Pont-Pioneer was the Technology Leader and provided the

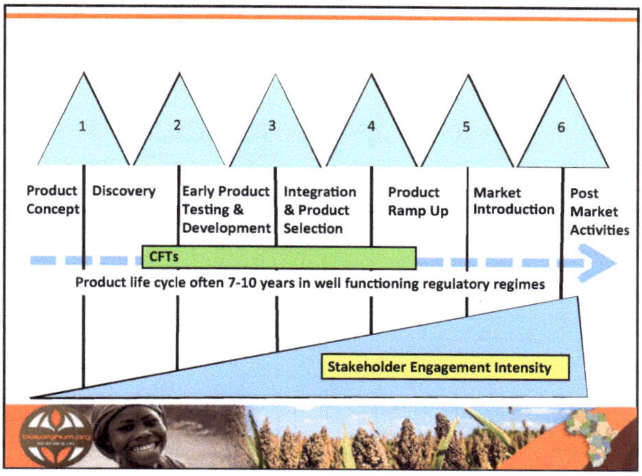

Diagram 13.1 General product development lifecycle (Credit: DuPont-Pioneer)

project PI, while the CSIR was the Technology Partner (through which technology transfer would get to African institutions). Based on this project leadership structure, a Project Communication Team (PCT), bringing together the communication teams from the three key organizations—Africa Harvest, Du Pont-Pioneer and the CSIR—was set up.

The communication strategy's goal was to keep project partners informed, engaged, and on message. It was also to address biosafety issues, as well as political challenges surrounding the application of permits to conduct contained greenhouse and field experiments. The process involved engaging key stakeholders such as regulators and government officials as part of the permit or policy approval process. The strategy also covered communication training and capacity-building—especially for scientists—within partner institutions.

The Important Role of CIMS in the ABS Project

The CIMS of a GM project is loaded with uncertainties. In addition to challenges related to technology development and performance, these projects carry a heavy load of uncertainty about regulatory approvals and the evolution of the policy climate, especially in developing countries. For the ABS Project, the implementation strategy required a CIMS that would ensure successful project implementation and final product acceptance. The project recognized that the ultimate test of success would be the uptake of the final products by farmers and consumers. In much of the intended target markets, these two categories overlapped considerably, creating challenges and opportunities.

As Diagram 13.1 shows, GM product have lengthy life cycles of between 7 and 10 years in well-functioning regulatory regimes. When it started, the ABS Project targeted five African countries, of which only three—South Africa, Burkina Faso, and Egypt—had regulatory systems that allow commercialization. Kenya has since

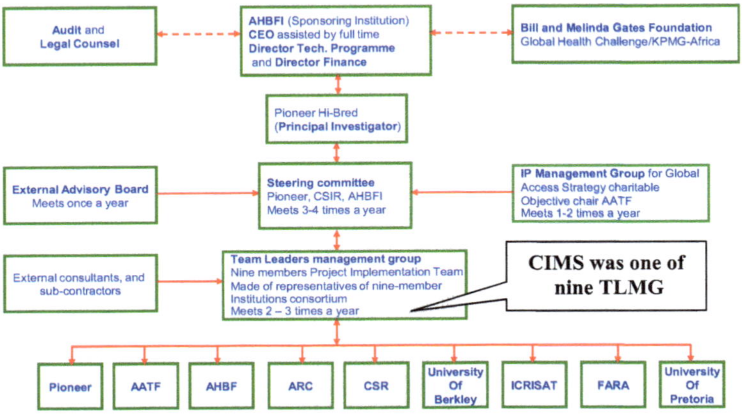

Diagram 13.2 ABS Organogram (Credit: Dr. James Onsando)

passed a law that allows commercialization while Nigeria's Biosafety Bill awaits Presidential assent. Given the challenging regulatory environment, the project's CIMS continues to take a long-term approach, with stakeholder engagement intensity increasing as the project progresses to its final goal.

Experience from similar projects (Acharya and Mackey 2008) shows that uptake of new crops by farmers is closely aligned to yield increase or reduced input costs. The ABS project is unique in that its most important benefit is improved nutrition. The CIMS therefore needed to be benefit-driven. Given the very long period between the research work and getting the product to farmers, the CIMS focused on the fact that the technology was designed to deliver the intended benefits without yield loss.

Diagram 13.2 is helpful in understanding the role of the CIMS within the ABS Project structure. As the contracting organization with the Bill and Melinda Gates Foundation (BMGF), Africa Harvest had the overall responsibility of the project. Pioneer, being the technology donor, provided the Principal Investigator (PI), while Africa Harvest Chief Executive Officer (CEO), was Co-PI. The project was governed by a Project Steering Committee (PSC), which consisted of the key organizations: Africa Harvest, Pioneer, and the CSIR. Operationally, the project had nine project implementation units, called the Team Leaders Management Groups (TLMG). The CIMS was one of the nine.

A significant shift happened soon after the project implementation started, forcing the CIMS to become one of the most important of the nine TLMG. The CIMS was severely tested and redefined when South Africa's regulatory body, the GMO Council, declined to issue the first permit application to grow ABS#1 in a (confined) greenhouse experiment. Although this issue will be discussed in detail later, it altered the strategic direction of the project in two ways. First, the project began to seriously explore the possibility of its experiments in other African countries and second, the CIMS was strengthened to support the new thinking.

In the project design, the CIMS focused on "benefit messaging." However, the realities in the ABS target countries required a shift—or de-emphasis—from farmers and consumers to scientists, regulators, and policymakers. CIMS core focus shifted to communication and issues management support for permit application for ABS events grown in the greenhouse trials (GHT) and confined field trials (CFT) in South Africa, Kenya, Burkina Faso, and Nigeria.

The CIMS ensured the project has required outreach materials for each stage in the biosafety and regulatory permit approval process. To ensure message consistency and effective deployment of project messages, project teams (and third party individuals and institutions) were trained during strategic periods related to permit applications. The project also reached out to critical stakeholders such as regulatory authorities and scientists (in National Agricultural Research Organizations as well as in Universities).

With regard to issues management, the CIMS provided support, especially at critical times, such as the period preceeding the GHT and the CFT. At appropriate times, various materials were developed and deployed to support specific project imperatives. For example, simplified policy briefs were used for biosafety and regulatory outreach, position papers were developed for targeted scientific symposia, and Question & Answers (Q&As) were used for media outreach.

To ensure consistency of information, the CIMS continuously framed issues as agreed by the project leadership, and ensured standardization across printed, online, and other information delivery platforms. To deliver agreed project messages and information at conferences, workshops, and other meetings, project scientists and third-party individuals and institutions were trained regularly. The project also shared information with clearly defined stakeholders to avoid "crossed wires". At critical times, spokespersons in different institutions were trained and agreement obtained on message content and effective delivery.

To keep the diverse institutions on track, the project developed an ABS Policy Manual, which documented what to do and what not to do in all areas, including communication and issues management. It spelt out the communication procedures that all consortium members would follow. For example, each partner institution had to designate a Communication Liaison Officer who would specifically coordinate all ABS project communication matters

CIMS Milestones and Roadmaps

The first 5-year phase (2005–2010) of the ABS Project had seven major objectives and over 100 milestones (see Diagram 13.3). The CIMS was one of the (major/project) objectives. It also had its own CIMS objectives, with each objective having specific milestones. Diagram 13.3 shows that during the first year, the CIMS had three objectives, nine activities and 15 milestones. Since the CIMS workplan was based on the overall project work plan, all activities and milestones were aligned to achieve the project goals.

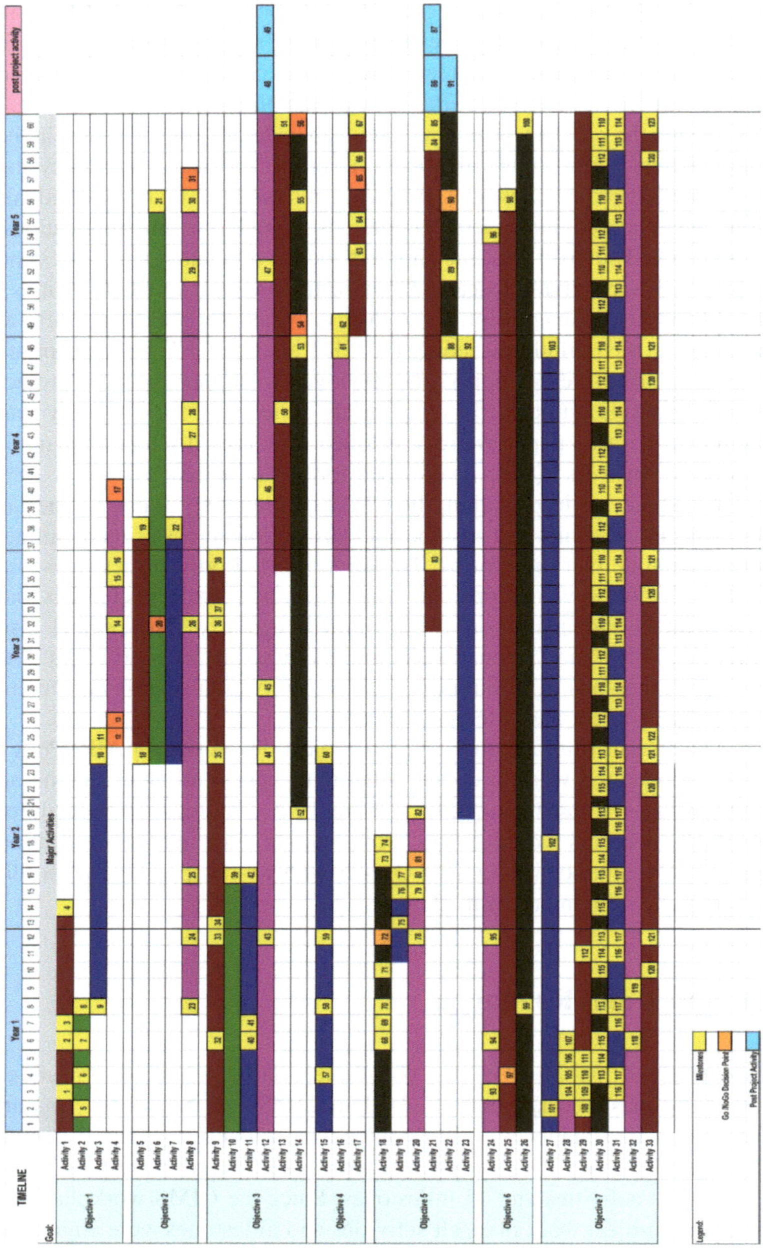

Diagram 13.3 ABS milestones

It was the ABS Project Manager's role to report progress to the funders. Since continued funding was linked to milestones, it was critical to achieve milestones or have a good explanation for any delays. The Project Manager regularly visited country teams and coordinated those tasked with oversight for the project's objectives. He also monitored go/no go milestones. These were critical milestones—such as the issuance of a permit to carry out certain experiments—that determined whether or not the project would move to the next phase.

As one of the nine TLMGs, the CIMS's decision-making unit (DMU) consisted of the communication directors of the three key organizations: Africa Harvest (Kenya), Du Pont-Pioneer (USA) and the South Africa-based Council for Scientific and Industrial Research (CSIR). It was chaired by Africa Harvest, and brought together all heads of communication from other project consortium members. The DMU ensured statements and messages were pre-approved, so as to facilitate speedy deployment across various channels as needed. Reporting to the PSC, the CIMS-DMU had the flexibility to deal with certain urgent and critical decisions, as long as it kept the PSC informed.

All the project's nine TLMGs had objectives, activities, and milestones that they were implementing. As Diagram 13.4 shows, during the first year of the project, the three major objectives for the CIMS were: (1) gather information related to the project, especially previous experience of similar projects, (2) develop a project white paper (on communication and issues management), and (3) finalize the media strategy, based on the information gathered.

Anti-GM Activism and Effect on the ABS Project

The CIMS was severely tested and redefined when South Africa's regulatory body, the GMO Council, declined to issue the first permit application to grow ABS#1 in a greenhouse experiment. Although the Council requested for clarity and additional information, matters were complicated when the issue reached the media before the project leadership were formally informed. This happened while the entire project leadership was in the US, making it very difficult—because of time differences – to manage the media interest and respond in a timely manner to enquiries.

The anti-GM organization, the African Centre for Biosafety (ACB) (Mayet 2007) was the first to publicly announce the decision by the regulatory authority, even before the permit applicants had been formally notified. "This decision was taken against the backdrop that Africa is the centre of origin for sorghum where (including in South Africa), a large number of sexually compatible weeds, wild relatives strains and races of cultivated sorghum occur," ACB stated.

ACB apparently disclosed the "real reason" for the Council's decision when they boasted of having lodged an objection to the ABS Project application "and raised strong concerns that GM sorghum would introgress into wild relatives". "Some activities just cannot be permitted and should be regarded as 'no-go' options," the organization said in its public statement. Long before the GMO Council released the reasons for the permit rejection, ACB said: "The risks posed by GM sorghum to

Goal: Pave way for product development and clear acceptance path for final product	
Objective 1:	*Gathering information related to the project, especially previous experiences of similar projects*
Activity 1:	**Define research questions specific to ABS1 (e.g food safety, environment, compliance)**
Milestones:	1. Consultation with consortium members and stakeholders
	2. Literature review with focus on similar projects
Activity 2:	**Define baseline and monitoring system (e.g. societal concerns, NGO concerns/support)**
Milestones:	1. Agree on baseline framework with consortium members
Activity 3:	**Confirm stakeholders buy-in to project (e.g. governments, African universities and agricultural organizations)**
Milestones:	1. Drawing up list of stakeholders and communicating project objectives
	2. Confirm stakeholders buy-in
Activity 4:	**Finalize media and PR outreach strategy**
Milestones:	1. Negotiation with African Economic Editors Forum (AEEF) and identify PR company(ies)
	2. Finalize draft media strategy framework
Objective 2:	*Develop Project White Paper*
Activity 5:	**Develop statements on key areas of focus (e.g.Global Access Strategy, Partnerships & Environment**
Milestones:	1. Agree with consortium members on key areas of focus
Activity 6:	**Stakeholder consultations and final signatures/buy-in to White Paper**
Milestones:	1. Negotiation with individual consortium members and stakeholders
	2. Workshops with consortium members and groups of stakeholders
Objective 3:	*Finalize Media Strategy*
Activity 7:	**Develop and agree on key message(s) concerning project**
Milestones:	1. Develop list of key messages
	2. Consotium members consultations and agreement on key messages
Activity 8:	**Finalize project publicity materials (print and electronic/website)**
Milestones:	1. Develop lists of materials required
	2. Finalize content, begin design and printing of materials
Activity 9:	**Agree on outsourced activities with AEEF and PR company**
Milestones:	1. Discussions & final agreements with AEEF and PR companies

Diagram 13.4 CIMS milestones and timelines

sorghum wild and weedy relatives cannot be tolerated at all, and the granting of a permit will be tantamount to a licence to contaminating Africa's heritage. Even containment in a level three facility will not negate the concerns that will remain, if the GM sorghum was to be tested in open field trials with the objective of commercialisation."

These were the same reasons the Council later cited. Dr Julian Jaftha, the then chairman of the GMO Council Genetically said "the council's main concern was an environmental one. There is a risk that modified sorghum might interbreed with its wild relatives in South Africa." (Jordan 2007). This was despite the fact that the application request was for *confined* greenhouse testing. Dr Gatsha Mazithulela, the then CSIR Executive Director of Biosciences, responded: "If we are not even allowed to grow genetically modified organisms in a greenhouse, which is a contained environment, then who is ever going to present evidence on what the risks are?" (Jordan 2007).

The then Minister for Land and Agriculture, Ms. Lulama Xingwana, praised the project's search for long-term solutions to the challenge of poverty and malnutrition. "Scientists should explore all possibilities of broadening the food base...and identify integrated approaches to maximise impact and adoption of new technologies." (Zvomuya 2007). What was not evident during the permit debacle was the fact that Africa was going through a period of intense anti-GM activism. For the third time, the SA regulatory authorities had also rejected Syngenta's application to grow GM maize in South Africa for the biofuel industry (Gosling 2007).

At around the same time, the former United Nations General-Secretary Kofi Annan was widely reported (Business Daily, Kenya, 17 July 2007) to have rejected GMOs (although he, and the organization he chaired, the Alliance for a Green Revolution in Africa (AGRA), clarified what they had said and intimated he had been misquoted). The alarming headlines—for example, *Kofi Annan says No to GM crops in Africa* or *Annan rules out the use of GMO's in the war on hunger in Africa*—emboldened anti-GM forces, and caused African regulatory authorities to be more cautious. One anti-GM organization even said "being an African, Kofi Annan fully understands the spiritual connection between the land, the seeds, and the ancestors." (http://www.flag-sa.org/blog/2007_07_01_archive.html)

Despite the apparent anti-GM activists' "victory", the South Africa permit rejection had a happy ending. After forming an independent panel to review the decision, the then Land and Agriculture Minister, Lulu Xingwana, overruled the GMO Council and approved the CSIR's application for greenhouse experiments on GM sorghum on the basis of "the potential scientific impact of the project in the long term." The CSIR immediately welcomed the ruling: "The decision is in the best interest of scientific inquiry, and provides a basis for making a difference to the neediest people of our continent," the then CSIR biosciences executive director Gatsha Mazithulela said.

The South African experience caused the project leadership to expand the number of countries in which the project experiments could be done. More specifically, permit applications were sought for various experiments in Kenya, Nigeria, and Burkina Faso.

Prioritizing Communication and Issues Management

The early challenges forced the project leadership to prioritize the CIMS. It was agreed that the vision of the CIMS would be "creating an enabling environment for all the partners for timely project implementation and product success". To achieve this vision, project partners commited themselves to ensure that all communication adhered to sound science, was transparent, and enhanced dialogue. The project also sought to ensure message consistency, regulatory compliance (voluntary and mandated), demonstrated safety, and focused on eventual product beneficiaries.

Diagram 13.5 ABS Project's communication approach

Recognizing that some of the project target countries lacked appropriate legislation to commercialize GM crops and products, the CIMS has a strong government relations bias, designed to support required biosafety and regulatory systems.

Among the communication challenges that helped in the design of the CIMS were:

- Creating and managing expectations related to the project among key target audiences
- Designing and deploying a differentiated communication strategy that still prioritized the key target audiences
- Proactive and reactive communication: the need to provide information on the project while responding to anti-GM activism
- Framing biotechnology as one of the tools and not the objective of the project

CIMS Conceptualization

Designing a CIMS that cut across nine institutions was a daunting task, given that each of the partners already had their institutional structures and communication strategies. The Diagram 13.5 reflects the strategy developed by identifying the best from each organization. To simplify the conceptualization process, six steps were developed. The *first step* involved agreeing on the ABS Project's "broad goal," which was "to develop a nutrition-enhanced sorghum product." The CIMS team saw the most critical measurement of performance as permit approval and

Decisions to Make	Audience 1	Audience 2	Audience 3
Audience Target:	• Regulatory authorities in 3 key target countries	Scientists in related disciplines	Media
Readiness:	Sharing Knowledge, Building Will & Reinforcing Action	Sharing Knowledge, Building Will & Reinforcing Action	Sharing Knowledge, Building Will & Reinforcing Action
Core Concerns:	Value: Want target countries to benefit from GM research Barrier: SA:Sorghum is not an important crop for South Africa KE: Over focuson maize BF: Sorghum too important	Value: Want science agriculture to thrive. Barrier: Fear that GM could "drift" and negatively impact science	Value: Want openness & information sharing. Barrier: ABS is based on outside, Western technology; Africa could be exploited

Diagram 13.6 How the CIMS made strategic choices

successful completion of confined greenhouse and field trials. To achieve this goal, the team would need to work closely with the technology group (within the project) and regulatory authorities (outside the project).

The *second step* involved "looking inwards" into the project to identify the assets and liabilities. The internal scan confirmed that the project had a diverse coalition of excellent scientists and professionals from other disciplines. Good communication skills also existed within the nine partner organizations. These were supported by various assets, such as websites and online newsletters, that could be used to deploy messages from the project. The biggest liability related to the size of the project, creating an enormous challenge of keeping all coalition members informed, engaged, and on the same message. There were also tight timeframes, and which objectives, activities, and milestones were likely to be affected by project dynamics during implementation.

The *third step* of CIMS conceptualization involved developing the appropriate messages for the project. This required either framing or re-framing the issues. Framing the issue was required in situations where there was nobody talking about the issue or the issue did not have prominence within target audiences. Reframing the issues was critical where the project desired to change the discussion and align it with communication goals.

Diagram 13.6 illustrates the framing/reframing exercise. In this particular case, we were strategizing on how to deal with issues around the decision by the GMO Council to deny the project a permit for greenhouse experiments. We identified three key audiences: regulatory agencies, scientists (in public and private sector), and the media. (Normally, we would not define the media as a target audience, but given the manner in which they were handling the issue, we realized the need to target them as an audience).

The *fourth step* was to develop appropriate messages. Diagram 13.6 shows that, to develop messages for these three audiences a good understanding of the core-concerns was needed. The audience readiness or receptiveness to the message

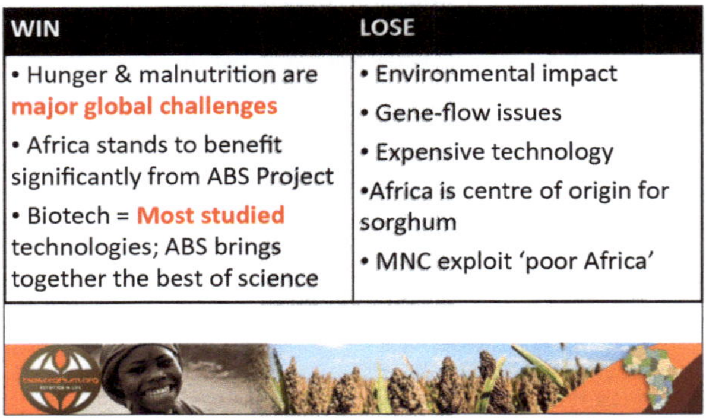

Diagram 13.7 Analyzing win–lose messages

would also influence how the message was framed. For example, from the decision made by the GMO Council, we deduced that sorghum was not a priority crop for South Africa and regulators were concerned about environmental issues. Although from a technology point-of-view, this reasoning is not sound (because the permit requested was for *confined* greenhouse experiments), outreach to this target audience would need to be sensitive to their perceptions. Understanding their core concerns and their readiness to engage with the project information helped us to better tailor our messages in a way that it achieved specific goals.

After the core concerns and readiness were established, the next challenge involved crafting the right messages. For example, we realized that at political and policy level, the messages resonated when we positioned the ABS Project as a possible solution to the major global challenge of hunger and malnutrition. For the African scientists—on whom the project was relying to convey the right messages—the issue was one of how they would be part and parcel of the research. For the regulators, the key issue was one that answered the question: has this been done elsewhere? A winning message therefore revolved around demonstrating that the ABS Project was relying on the best science and, perhaps, the most researched agricultural technology in the world (Diagram 13.7).

The *fifth step* involved measuring the success of the CIMS. Success was closely linked to permit approvals and successful completion of confined greenhouse and field trials. We also looked at the materials developed and their alignment to project goals, as well as outreach through newsletters and the media. Linked to the fifth step was the *Sixth Step*, which involved a "reality check" or evaluation of whether the strategy was working. This involved bringing together the team and asking the following questions:

- Was the strategy doable?
- Were resources in line with strategy?
- Did internal and external scan support the decisions made?
- Did tactics move us towards our objective, and did we reach the appropriate audience(s)?

- Did we obtain (internal and external) buy-in to implement project in a timely fashion?
- Can we measure progress continuously?

Conclusion, Building on Current Success and Recommendations for the Future

From a technology point of view, the project's first phase success included:

1. Optimization and improvement of sorghum transformation systems, leading to a significant increase in the sorghum transformation efficiencies. This provides a global opportunity for additional improvement of the sorghum crop through genetic engineering;
2. Developing the world's first "golden sorghum" with enhanced levels of pro-vitamin A, reduced phytate, and an improved protein profile. Pro-vitamin A amounts were within the range of those obtained from the Golden Rice Project;
3. Bioavailability studies showed increased rates of zinc and iron absorption;
4. Successful field and greenhouse trials. At the end of Phase 1, six field trials in the USA had been done, and greenhouse trials had taken place in Kenya and South Africa. (At the time of writing the paper, several seasons of CFTs have been undertaken in Kenya and Nigeria);
5. The Intellectual Property audit for freedom to operate status has been achieved for all the genes used in the ABS project in all target countries and regions in Africa;
6. Capacity building and infrastructural development has been undertaken in the USA for African scientists in partnering institutions from countries of deployment in genetic transformation, throughput breeding, biosafety, and regulatory;
7. ABS traits have been backcrossed to popular African sorghum varieties, and the traits have shown stability in agronomic and ABS traits in African varieties (including Marcia, Tegemeo, KARI Mtama I, Sundanse, Malisor) laying the foundation for the future.

The success of the project also confirms the success of the Communication and Issues Management Strategy (CIMS). After the end of Phase 1, the project's limited funding required narrower focus for all project areas. Like many similar biotech projects in Africa, funding challenges mean that Africa remains "open game" for anti-GM groups. Placed against staid, scientific explanation, anti-GM messages have found a ready audience because they are designed to play to existing misconceptions. These messages are well-received because citizens and farmers alike are not aware of other crop-improvement technologies that preceded modern biotechnology.

While increased funding for biotech communication would make a difference, the immediate solution seems to lie in African biotech projects combining forces. A united, African biotech communication platform would create the required force, reduce duplication, and unlock greater value through synergistic communication strategies. Such a platform would bring together key biotech proponents, and build bridges with allies and stakeholders to ensure Africa truly speaks with one voice on biotechnology.

The platform would provide accurate information on each project and, as appropriate, push for the general acceptance of the technology. It would jointly address issues that cut across the projects while allowing each project to deal with project-specific issues. Sharing and segmenting stakeholders would improve general awareness and bring greater effectiveness in targeting niche audiences, addressing emerging issues, and responding quickly, but uniformly, during times of crisis.

Beyond Talking, Demonstrating the Benefits of the Technology

The experience of the ABS Project confirms that uptake of biotechnology and future success is intricately tied to demonstrated economic benefits and a neutral or positive impact of the environment. Despite the challenges in obtaining required permits for lab, contained, and open field trials, these are important, especially in developing appropriate polices and legislation.

If the next generation of biotech crops with superior traits, improved properties, and quality traits are to be created and deployed in developing countries, then Africa, Asia, and Latin America must be encouraged and supported, not just to be markets but developers of the technology. Anti-biotech forces know that preventing the lab and field trials is only the first line of attack; the battle will be determined by showcasing this technology. The two misconceptions—GM as unnatural and the presence of biotech crop material as detrimental to other crops (McHughen and Wager 2010)—can only be put to rest by allowing continents such as Africa to carry out their own experiments.

References

Acharya SS, Mackey GM (2008) Socio-economic impact assessment of the Tissue Culture Banana Industry in Kenya. http://issuu.com/africaharvest/docs/socio-economic

Allen K (2013) Meet Calestous Juma, Africa's genetically modified crop 'optimist'. www.thestar.com 17 June. http://www.thestar.com/news/world/2013/06/17/meet_calestous_juma_africas_genetically_modified_crop_optimist.html

De Villiers J (2013) Chairman of Grain SA. How SA will deal with food crisis. Financial Mail, 8 Aug 2013. http://www.fm.co.za/fm/Features/2013/08/08/how-sa-will-deal-with-food-crisis?service = print
FAO (2012) World Agriculture: towards 2015/2030. Summary report. ftp://ftp.fao.org/docrep/fao/004/y3557e/y3557e.pdf
Gedye L (2004) Row over 'mutant' Aids drug. Mail and Guardian, South Africa 23 July. http://mg.co.za/article/2004-07-23-row-over-mutant-aids-drug
Gosling M (2007) Government rejects 'unsafe' modified maize. The Star, South Africa, 28 March. http://www.iol.co.za/news/south-africa/government-rejects-unsafe-modified-maize-1.320722
Jordan B (2007) SA nixes super sorghum. Sunday Times, South Africa, 9 Sept
Kikulwe E, Wesseler J, Falck-Zepeda J (2010) Consumer perceptions towards introducing a genetically modified banana (*musa* spp.) in Uganda. Acta Hort 879:175–183, http://www.actahort.org/books/879/879_16.htm
Mayet M (2007) Africa's sorghum saved: applause for second GM sorghum rejection, 3 Feb. http://www.acbio.org.za/index.php/media/64-media-releases/12-africas-sorghum-saved-applause-for-second-gm-sorghum-rejection
McHughen A, Wager R (2010) Popular misconceptions: agricultural biotechnology. New Biotechnol 27(6):724–728
Namibia Economist, Windhoek, Namibia (2011) GM crops contribute to poverty reduction. http://www.thefreelibrary.com/GM+crops+contribute+to+poverty+reduction.-a0258357516
www.allafrica.com 11 March 2009. Nigeria: environmentalists call for end to GM cassava trials http://allafrica.com/stories/200903130668.html
Zietz J, Seals A (2006) Genetically modified maize, biodiversity, and subsistence farming in Mexico. http://capone.mtsu.edu/berc/working/Genetically%20modified%20maize%20-WP.pdf
Zvomuya F (2007) Supersorghum: no appeal yet. Farmers Weekly, South Africa, 2 March

Chapter 14
Social Audits and Their Role in Stakeholder-Oriented Innovation and Fostering Accountability and Trust in Agricultural Biotechnology Development Programs

Obidimma C. Ezezika and Jessica Oh

Abstract Major efforts are needed to improve the efficiency of the agricultural sector in sub-Saharan Africa, for which it is an important engine of economic development. Advances in agricultural biotechnology (agbiotech) have led to the development of crops that have the potential to enhance the agricultural productivity of resource-poor smallholder farmers. Most of the research and development of these biotech crops are led by public–private partnerships (P3s). However, such partnerships face challenges to accountability and trust due to public distrust of the private sector, and the controversy surrounding the application of biotechnology to agriculture. There is also the concern about ensuring that these projects effectively engage in stakeholder- and farmer-oriented innovation. This chapter explains how social audits can play a role in fostering accountability and trust, which are important components in the effectiveness of P3s. It also shows the importance of social audits in engaging farmers and stakeholders, and providing them with a voice in the innovation process.

Keywords Stakeholder engagement • Innovation • Farmer • Social audit • Trust • Accountability • Public-private partnership

---

O.C. Ezezika (✉)
MaRS Centre, South Tower, Sandra Rotman Centre, University Health Network and University of Toronto, Suite 406, 101 College Street, Toronto, ON M5G 1L7, Canada

Dalla Lana School of Public Health, University of Toronto, Toronto, Canada

African Center for Innovation and Leadership Development,1938 Bloor Street, PO Box 30007, High Park, Toronto, Ontario M6P 4J2, Canada

National Biotechnology Development Agency, Federal Ministry of Science and Technology, Nigeria
e-mail: obidimma.ezezika@acild.org

J. Oh
MaRS Centre, South Tower, Sandra Rotman Centre, University Health Network and University of Toronto, Suite 406, 101 College Street, Toronto, ON M5G 1L7, Canada

African Center for Innovation and Leadership Development,1938 Bloor Street, PO Box 30007, High Park, Toronto, Ontario M6P 4J2, Canada

F. Wambugu and D. Kamanga (eds.), *Biotechnology in Africa*, Science Policy Reports 7, DOI 10.1007/978-3-319-04001-1_14,

Abbreviations

AATF	African Agricultural Technology Foundation
agbiotech	Agricultural biotechnology
CIET	Community Information and Epidemiological Technologies
CIMMYT	International Maize and Wheat Improvement Center
GM	Genetically modified (not expanded in text, no need)
NGO	Non-governmental organization
P3	Public–private partnership
PI	Principal Investigator
R&D	Research and development
SSA	Sub-Saharan Africa
WEMA	Water Efficient Maize for Africa

Introduction

Agricultural biotechnology is increasingly being recognized as a tool for increasing food production and improving food security in sub-Saharan Africa (SSA), and public–private partnerships (P3s) are playing an important role in the process. P3s are often clusters of local and foreign partners that form both formal and informal coalitions and consortia to handle an array of challenges and opportunities that neither the private nor public sector can address independently (Hall 2006). P3s are important for the development of agricultural biotechnology due to the private sector's expertise in technical capacity and the public sector's local knowledge and experience in breeding and germplasm.

In the case of agricultural biotechnology, such partnerships involve collaboration of local agricultural research institutes, universities, multinational seed companies, and funders. It has therefore been recognized that the public and private sectors need to capitalize on mutual strengths to accelerate the process of development and field deployment of technologies for the benefit of resource-poor farmers. Partnerships between the private sector—which has proven the capacity to bring technologies to farmers (in the form of seeds and other agricultural inputs)—and the public sector—which has the capacity for agricultural research and local expertise and knowledge—are therefore central to attaining the agricultural goals of food sufficiency and improved livelihoods for people in Africa.

An example of such a P3 is the Water Efficient Maize for Africa (WEMA) Project, funded by the Bill & Melinda Gates Foundation and the Howard G. Buffet Foundation. It is operated by the African Agricultural Technology Foundation (AATF) in partnership with Monsanto, the International Maize and Wheat Improvement Center (CIMMYT), and the National Agricultural Research Systems in five SSA countries—Kenya, Mozambique, South Africa, Tanzania, and Uganda. The Project's goal is to deliver royalty-free, drought-tolerant maize varieties to smallholder farmers in SSA to increase agricultural productivity and protect them from the risks posed by drought (AATF 2013).

Challenges to the Effectiveness of P3s

While these partnerships have the potential to respond to wide-ranging global challenges by combining resources and expertise, experience has shown that success depends on meticulous handling of subtle and potentially fragile legal, confidential, ethical, and sociocultural elements. Partnerships such as WEMA can be undermined by the controversy surrounding genetic modification in agricultural development—thereby making actual delivery of the technology difficult and adoption of the end product less likely. Agricultural biotechnology is mired in much doubt, fear, and conflicting understandings—all of which are exacerbated by the added complexity of P3s. In particular, there exists mistrust and mutually negative perceptions among partners, which stem from cultural and ideological differences—both real and perceived—which have been identified as a primary impediment to effective partnership (Spielman and Grebmer 2004; Spielman and Grebmer 2006). There is mistrust among private and public sector stakeholders about each other's motives and capabilities in carrying out humanitarian projects, which is a major hurdle to successful partnerships between the two sectors.

Even when an agbiotech P3 functions well internally, the project can still be stymied by the skepticism of, and oftentimes opposition from, the community in which it plans to carry out its activities. Stakeholders' distrust in the private sector, particularly multinational agricultural biotechnology companies, contributes to their resistance toward adopting new technologies (Ezezika et al. 2012a). The public's fear about corporate control of agriculture is not unwarranted, as the global seed industry is currently dominated by only a few transnational corporations (Howard 2009). We found that farmers therefore fear losing control over traditional farming practices such as seed sharing and storage, which would lead to reliance on private companies for seeds and eventually render traditional crops extinct (Ezezika et al. 2012a).

Poor communication about GM crops was also found to be a major factor hindering adoption of agricultural biotechnology. In particular, limited public understanding of GM technology, coupled with inaccurate portrayal of GM foods by anti-GM groups and alarmist media reporting, serve to skew public perception about agricultural biotechnology (Ezezika et al. 2012a).

What Is a Social Audit?

Social audits can play an effective role in mitigating some of the risks that are closely linked to P3s working on projects developing new agricultural technologies, some of which are controversial due to the involvement of genetic modification. Social auditing can be defined as "a process whereby an independent audit team collects, analyses, and interprets descriptive, quantitative, and qualitative information from stakeholders to produce an account of a project's ethical, social, cultural, and commercialization performance and impact" (Ezezika et al. 2009). Particularly, social audits can help create a culture of trust, which is essential for partners to be able to work together and also gain the acceptance of the intended beneficiaries of their project. Social auditing is considered an important tool for building trust (Gao and Zhang 2006), as well as for improving accountability and transparency (Ebrahim 2003; O'Dwyer 2005; Zadek and Raynard 1995).

Social auditing has been practised since the 1970s, with Abt Associates, a US consultancy firm, being one of the first to incorporate social audit accounts into its annual report (Abt Associates 1976). By the 1990s, social auditing took a more systematic approach, and had become a practice for which the goal would be to improve the transparency and accountability of organizations and re-orient their activities towards the interests of their stakeholders (Zadek and Raynard 1995). Non-governmental organizations (NGOs) and social enterprises have also practised social auditing (Ebrahim 2003). In 2008, for example, the academic NGO Community Information and Epidemiological Technologies (CIET) conducted a community-based social audit in two districts in Afghanistan to document experiences and views of health services from the perspectives of individual households, with the aim of raising the quality of health services and minimizing inefficiencies (Cockcroft et al. 2011). Private companies and businesses have also applied the practice of social auditing. For example, Britain's media company Guardian News & Media has adopted a social audit system to assess how its operations meet its core ethical values and verify the honesty and completeness of its sustainability reporting. The result of this system has been a comprehensive sustainability effort that has been communicated transparently to stakeholders. Readers, in turn, have "expressed high levels of trust in the Guardian's brand of journalism" (Jaehnig and Onyebadi 2011).

Social auditing can be likened to financial auditing, in that performance data are collected and then reviewed by an independent and external auditor who verifies that the information is accurate. The auditor issues a statement confirming the accurate representation of the business or project. Social auditing and financial auditing thus employ, in essence, the same practices, but the former involves reporting on how resources are used to deliver social outcomes. Most importantly, social audits make an organization or program accountable for the social objectives it declares or to which it has committed itself. Social auditing is also relatively new, with less developed methods—whereas financial auditing employs long-established methods and widely accepted principles.

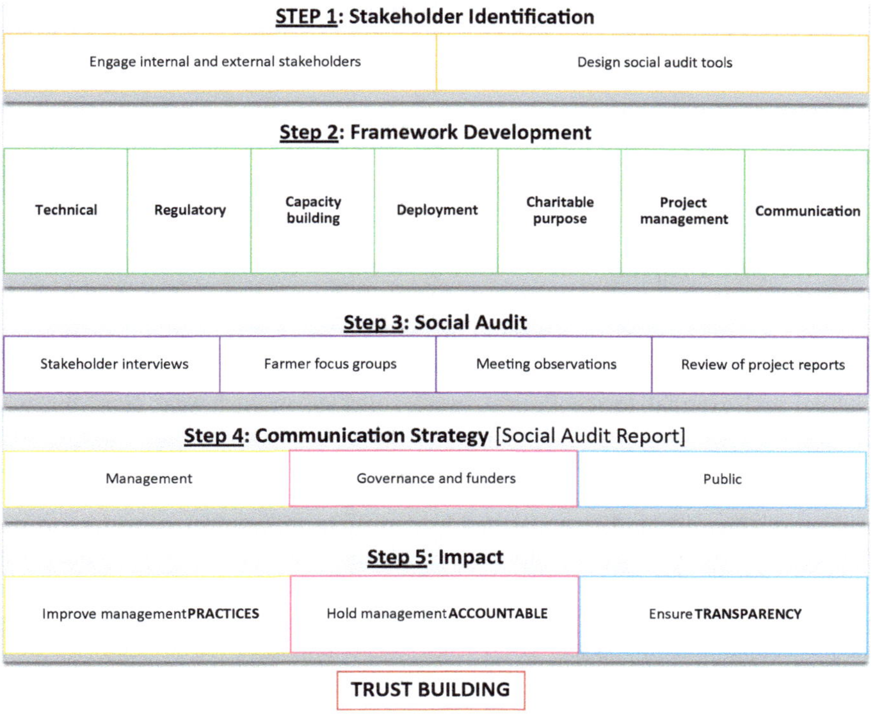

Fig. 14.1 Social audit model applied to the WEMA Project

A Social Audit Model for an Agbiotech P3

In response to the challenges faced by agbiotech P3s, we created a social audit model (Fig. 14.1), incorporating feedback from several project stakeholders, to apply to the WEMA project. While the concept of social auditing is not new, the social audit model we designed differs from other popular models in three major ways. First, the model is tailored to projects operated by P3s. Second, the model aims to improve internal management of the project and strengthen its public accountability—not one or the other exclusively. Third, the model is based primarily on stakeholder consultation and engagement.

We created a framework that took into account the goals of the WEMA Project, which are divided into seven major components: technical, regulatory, capacity-building, deployment, charitable purpose, project management and governance, and communication (Ezezika et al. 2009). We refer to these components as "audit lenses" (shown in Fig. 14.1), which shape the following four processes through which ethical, social, and cultural issues are made explicit: 1) interviews with stakeholders, 2) focus groups with farmers, 3) meeting observations, and 4) review of project reports. The model is premised on a 1-year project cycle in which an account of ethical, social, and cultural issues is produced. The results and recommendations of the account are fed back into the next cycle through the funders and the Principal Investigators (PIs) of the project.

Applying the Social Audit Model

The WEMA Project

The model was applied to the WEMA Project over 5 consecutive years starting in 2008 to evaluate the Project's performance. Views of 100 project stakeholders, internal and external to the Project, were collected in each year from the five participating countries (Kenya, Mozambique, Tanzania, South Africa, and Uganda). Stakeholder perspectives represented a wide range of stakeholder groups such as farmers, national agricultural research institutes, seed companies, and regulators (Fig. 14.2). These stakeholders' trust in the project is critical to the project's ability to meet its humanitarian goals. Selection of stakeholders was done via snowball sampling and based on stakeholders' knowledge of the WEMA Project, to ensure they were capable of responding to questions related to the project partnership, project governance, and communication strategy (see Box. 14.1 for sample questions).

Box. 14.1

Sample questions from the social audit questionnaire

1. What ethical, social and/or cultural issues have been encountered in the technical work of the project?
2. Has sufficient preparation been made by the WEMA partners to address these issues if and when they arise? Please explain briefly.
3. What would you consider to be the important concerns and benefits of this project to (respective stakeholder group)?
4. Which of these concerns/benefits are being considered implicitly or explicitly in project planning and implementation? Please explain briefly.
5. What potential commercialization issues do you foresee in the WEMA Project, and have preparations been made by the WEMA partners to effectively address these issues if and when they arise?
6. Do you see any difficulties that may arise as this project advances? If so, how should they be addressed?

Findings from the social audit were reported to project management, funders, and stakeholders. The WEMA project teams were also provided with recommendations in the report, which comprise action plans set out to ensure that the issues

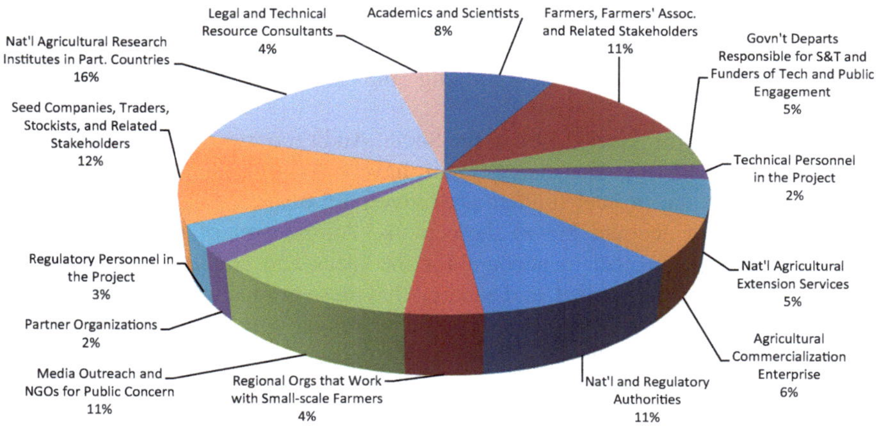

Fig. 14.2 Distribution of stakeholders in the 2012 social audit. $N = 100$

raised by stakeholders were addressed in the next project cycle and that transparency, accountability, and management practices were improved. Some of the key findings reported include:

- Overall evaluation of the WEMA Project is 'good', and varies among stakeholder groups.
- Communication with stakeholders has improved since the preliminary audit.
- Stakeholders want transparency and input on the potential characteristics of WEMA maize.
- There is a perceived need for capacity-building of national agricultural research and regulatory systems.

Greater detail of the findings from the Social Audits of the WEMA Project (alongside WEMA management's responses to the reports) can be found in the Social Audit Reports posted on the AATF website (AATF 2012).

The Impact on Building Trust

Action steps, in line with key findings from the social audit, have been actively integrated into project plans by multiple project teams in each year the social audit has been implemented. The WEMA Project management has acknowledged, and actively incorporated, the issues identified and recommendations made in the Social Audit reports into team work plans. Some of these plans have included strategies on communication with stakeholders, project transparency, and stakeholder input, as well as awareness building of the project's charitable purpose through clear communication of intellectual property rights structure (AATF 2013). Building capacity through improved regulatory approval and closer integration of national agricultural research systems into the operations committee were also taken up by the WEMA management, to foster trust with a diverse group of stakeholders and maximize the benefits of incorporating the expertise of locally engaged organizations.

Public reporting of issues raised in the Social Audit reports, and responses to the issues by the WEMA Project management, have helped to create transparency in the project. Both the project funders and governing board of the AATF held the management team accountable to these plans. Application of the Social Audit Model enabled the project to account for the ideas and concerns of all parties involved in the WEMA Project, and ensured that these groups were informed about the issues that arose and how they would be addressed in the Project. The combination of accounting for the viewpoints of all parties, and disseminating the key findings of this information to all parties, allowed for transparency and accountability, and helped to align the goals and interests of the various parties, which otherwise might not have been openly communicated and negotiated.

Project stakeholders also responded favorably to their involvement in the social auditing process. In their feedback, they expressed appreciation for the "professional approach" of the social audit, and indicated that it expanded their knowledge of the project. Stakeholders also appreciated that their perspectives on the project were accurately reflected in the Social Audit reports, and were grateful for how the Social Audits provided openness or transparency in the project. They expressed interest in seeing Social Audits conducted on other projects led by the AATF, and indicated that they would recommend the social auditing service to other technology development projects.

The Importance of Stakeholder Engagement in Innovation

Stakeholder-Oriented Innovation Through Social Audits

Social audits revolve around stakeholder engagement. We placed most emphasis on the importance of engaging the *stakeholder*—defined as a person or organization that has an interest, or has invested resources, in a given project or organization (UN PAN 2005). From 5 years of experience with the WEMA Project, perhaps the greatest lesson learnt is the importance of 'embedding' the stakeholder in the innovation process—or what we refer to as "stakeholder-oriented innovation." For example, the importance of stakeholder-oriented innovation emerged in one of the findings in the 2009 Social Audit regarding the issue of 'stacking traits'—the incorporation of multiple GM traits in a single variety of a crop. We derived two key lessons. First, we found that stakeholders of the WEMA Project wanted concerns about stacked traits to be publicly and transparently addressed; this request stemmed from fear that the project could be a Trojan horse, promising further traits in a single crop and therefore extending the GM market base of large agbiotech corporations. Second, with farmers increasingly recognizing the benefits of using crops with stacked traits, we found that technologies that do not contain important characteristics preferred by farmers are less likely to be adopted. These findings from the social audit were eventually converted into actions taken by the WEMA Project. This is an example of stakeholder-orientated innovation.

The identification of various challenges and their causes through the social auditing of the WEMA Project exemplifies how a viable technology grounded in humanitarian intentions can fail if it does not consider the various ethical, social, and cultural issues obstructing delivery of the new technologies to farmers. Good intentions, sound science, and rigorous research are insufficient for successful delivery of technologies to the people for whom they are intended. Those who are developing and delivering the technologies must be cognizant of the fact that they are dealing with human subjects—that the recipient and beneficiary of their product is a social being. The potential solutions to working with these subjects are therefore as numerous and diverse as the needs they aim to address; the complexity of human behavior and diversity of perceptions must be considered. Even the most accurate and comprehensive of financial audits can only do little in disclosing the actual performance of a given project, program, or organization vis-à-vis its declared core values, because they do not employ the kinds of systematic measurements that assess the impact of non-financial objectives (UN PAN 2005). Social audits can play more than merely a supplemental role in uncovering the diverse needs and concerns of the intended beneficiaries and revealing the various issues that must be addressed to ensure that the project's goals are met.

Despite the many promising innovations in the field of agricultural development that are now within reach, the only type of innovation that has been shown to make a difference and ultimately succeed is that which is culture- and context-specific and built with the needs of stakeholders in mind. Agricultural innovation, especially that involving biotechnology, is not a one-way process from the laboratory to the Ministry of Agriculture to the farms and the farmer; rather, it is an interactive, social process that is fueled by the engagement of various players and the feedback they provide.

Stakeholder Engagement in Agbiotech P3s in Africa

Working with the WEMA Project led us to explore whether there were other P3s in Africa working on developing and deploying biotech crops, and how these partnerships engaged with their stakeholders. We therefore carried out a series of case studies[1] on agbiotech P3s in Africa—specifically in Burkina Faso, Egypt, South

[1] Full access to the case studies is available at: http://www.agricultureandfoodsecurity.com/supplements/1/S1

Africa, Kenya, Nigeria, Tanzania, and Uganda. These case studies culminated in a publication in the peer-reviewed journal *Agriculture & Food Security* as a special supplemental issue.

In Burkina Faso, for example, we found that a lack of sufficient community engagement during the research, development, and implementation phases of the project in the country created challenges to building trust in the partnerships (Ezezika et al. 2012b). Specifically, a significant challenge arose from lack of communication among researchers, journalists, and the community. The disconnect between researchers' knowledge of the technology and the uninformed community fostered public distrust in the technology and the research and development (R&D) process, thereby hindering further outreach efforts by the project to alleviate concerns of a skeptical and apprehensive public.

The Place for Social Audits in the Future of Biotech Crops

Since the later half of the twentieth century, African countries have been experiencing increasing agricultural risks caused by drought, soil erosion, pests, and diseases. This is a serious concern, since an estimated 80 % of the population in Africa depends directly or indirectly on agriculture for their livelihood. In an effort to capitalize on the potential benefits of modern biotechnology while ensuring protection from potential risks, most African countries have already signed and ratified the Convention on Biological Diversity as well as the Cartagena Protocol on Biosafety (Nang'ayo 2006). Burkina Faso, Egypt, and South Africa—the three African countries that allow for commercialization of GM crops—are each seeing increases in biotech hectarage year after year (James 2011). Yet, these growing numbers, as well as the growing recognition of agricultural biotechnology as a tool for improving food security, are being followed by a proportional increase in fear and skepticism about the safety and public health impact of biotech crops—and not surprisingly so, considering the anti-GM movements taking place worldwide. It is thus essential that biotech crops are created with both the end-user and local communities in mind—by addressing ethical, social, cultural, and commercialization issues as they move from the laboratory to the village. Social audits can greatly facilitate this delivery by making use of tools and techniques that are aimed at systematically weaving the community's voice into the evaluation of a given project's impact. The result is an increase in the community's trust in such projects, and clear direction on how, and where, improvements can be made in order to better serve the intended beneficiaries.

Conclusion

The Social Audit Model has helped the WEMA Project to effectively build trust among the project partners and between the project and the public. In turn, contributions have been made to mitigating risks associated with lack of trust in, and within, the WEMA Project to ensure effectiveness in the governance of the project and that the project's humanitarian goals are met. The increase in the number of P3s working on developing biotech crops signal a special role that social audits can play in maximizing a project's effectiveness and social impact. From working and interacting with the WEMA Project, we have increasingly realized that social audits need to be integrated as a component in projects developing GM technologies, as such projects must be accountable, not only to those who fund them, but to the intended beneficiaries—the community—as well. Ultimately, the community's voice will be critical to ensuring the effectiveness of the project.

Acknowledgements This work was funded by the Bill & Melinda Gates Foundation, and supported by the Sandra Rotman Centre, an academic centre at the University Health Network and University of Toronto. The findings and conclusions contained within are those of the author, and do not necessarily reflect official positions or policies of the foundation. We would like to thank Jennifer Deadman and Justin Mabeya for helpful comments on earlier drafts of the manuscript.

References

AATF (2012) Water-efficient maize for africa (WEMA): Audit Reports. Retrieved 18 Feb 2013, from http://wema.aatf-africa.org/audit-reports

AATF (2013) Water-efficient maize for africa (WEMA). Retrieved 18 Feb 2013, from http://beta.aatf-africa.org/projects/aatf_projects//wema/

Abt Associates (1976) Annual report and social audit. Abt Associates, USA

Cockcroft A, Khan A, Ansari NM, Omer K, Hamel C, Andersson N (2011) Does contracting health care in Afghanistan work? Public and service-users' perceptions and experience. BMC Health Serv Res 11(Suppl 2):S11

Ebrahim A (2003) Accountability in practice: mechanisms for NGOs. World Dev 31(5):813–829

Ezezika OC, Thomas F, Lavery JV, Daar AS, Singer PA (2009) A social audit model for agro-biotechnology initiatives in developing countries: accounting for ethical, social, cultural and commercialization issues. J Technol Manage Innov 4(3):24–33

Ezezika OC, Daar AS, Barber K, Mabeya J, Thomas F, Deadman J, Wang D, Singer PA (2012a) Factors affecting agbiotech adoption in sub-Saharan Africa. Nat Biotechnol 30:38–40

Ezezika OC, Barber K, Daar AS (2012b) The value of trust in biotech crop development: a case study of Bt cotton in Burkina Faso. Agric Food Secur 1(Suppl 1):S2

Gao SS, Zhang JJ (2006) Stakeholder engagement, social auditing and corporate sustainability. Bus Process Manage J 12(6):722–740

Hall A (2006) Public–private sector partnerships in an agricultural system of innovation: concepts and challenges. Int J Technol Manage Sustain Dev 5(1):3–20

Howard PH (2009) Visualizing consolidation in the global seed industry: 1996–2008. Sustainability 1:1266–1287

Jaehnig WB, Onyebadi U (2011) Social audits as media watchdogging. J Mass Media Ethics 26:2–20
James C (2011) Global status of commercialized Biotech/GM crops: 2011. (ISAAA Brief No. 43). ISAAA, Ithaca
Nang'ayo F (2006) The status of regulations for genetically modified crops in countries of sub-Saharan Africa. African Agricultural Technology Foundation, Nairobi
O'Dwyer B (2005) The construction of a social account: a case study in an overseas aid agency. Account Organ Soc 30(3):279–296
Spielman DJ, Grebmer K (2004) Public–private partnerships in agricultural research: an analysis of challenges facing industry and the Consultative Group on International Agricultural Research. EPTD Discussion Paper No. 113
Spielman DJ, Grebmer K (2006) Public–private partnerships in international agricultural research: an analysis of constraints. J Technol Transf 31:291–300
United Nations Public Administration Network (2005) Social audit: a toolkit: a guide for performance improvement and outcome measurement. Centre for Good Governance, Hyderabad
Zadek S, Raynard P (1995) Accounting works: a comparative review of contemporary approaches to social and ethical accounting. Account Forum 19(2/3):164–175

Part VII
Political Challenges

Chapter 15
Does Africa Need Political Will to Overcome Impediments to GM Crop Biotechnology Applications for Agricultural Economic Growth and Development, as in the Case of Brazil, Argentina, and India?

Walter S. Alhassan and Adewale A. Adekunle

Abstract Sub-Saharan Africa (SSA) is a chronically food-insecure region, exacerbated by increasing population and the resulting decline in the size of arable land holdings, declining soil fertility, intractable pests and diseases, climate change phenomena (drought and floods), social conflicts, and the lack of a generally enabling policy environment for agriculture.

Modern biotechnology breakthroughs can be harnessed to address many of the intractable biotic (pests and diseases) and abiotic stresses (soil fertility declines) of modern agriculture.

Genetically modified (GM) crop production has seen a phenomenal 94-fold increase from 1.7 M ha in 1996 to 160 M ha in 2011. In 2011, Brazil (30.3 M ha), Argentina (23.7 M ha), and India (10.6 M ha), in that order, are the largest GM crop-producing countries next to USA (69.0 M ha) in the world. These countries have reaped immense financial and environmental benefits from the adoption of GM technologies in their agriculture.

This paper examines the status of GM product deployment and its impact on the economies of these countries, and in Burkina Faso, Egypt, and South Africa. The nature of political will demonstrated by these countries could serve as lessons for African countries, and spur the development and application of modern biotechnology in their agriculture on a need basis.

Keywords Africa • GM crop • Political will

W.S. Alhassan
Biotechnology Consultant, Forum for Agricultural Research in Africa, Accra, Ghana
e-mail: walteralhassan@hotmail.com

A.A. Adekunle (✉)
Director, Partnerships and Strategic Alliance, Forum for Agricultural Research in Africa, Accra, Ghana
e-mail: aadekunle@fara-africa.org

F. Wambugu and D. Kamanga (eds.), *Biotechnology in Africa*,
Science Policy Reports 7, DOI 10.1007/978-3-319-04001-1_15,

Abbreviations

Bt	*Bacillus thuringiensis*
BT/HT	Two-trait stacked gene variety of crop (*Bacillus thuringiensis* insect resistance, and herbicide tolerance)
CPB	Cartagena Protocol on Biosafety
CTNBIO	Brazilian National Technical Commission on Biosafety
CTNBIO	Brazilian National Technical Commission on Biosafety
DOST	Department of Science and Technology (the Philippines)
EMBRAPA	Brazilian Agricultural Research Corporation
GE	Genetically engineered
GM	Genetically modified
GMO	Genetically modified organism
HT	Herbicide-tolerant
IRRI	International Rice Research Institute
ISAAA	International Institute for the Acquisition of Agri-Biotech Applications
LMO	Living modified organism
MAS	Marker-assisted selection
NCBP	National Committee on Biosafety in the Philippines
PRSV	Papaya ringspot virus
SSA	Sub-Saharan Africa
VR	Virus-resistant

Introduction

Approximately 868 million people (12 % of the world's population) are undernourished (FAO 2012). Of these, 234 million (27 % of the undernourished in the world) live in sub-Saharan Africa (SSA). One out of every four of the 840 million people living in SSA lack sufficient food to live a healthy life (FAO 2012).

Over 50 million African children are suffering from chronic malnutrition, and 40 % of women are malnourished. Each year, 60 % of the under-5 and 50 % of maternal mortalities are due to malnutrition (Gyase 2011).

The food-insecure situation of SSA is exacerbated by increasing population and the resulting decline in the size of arable land holdings, declining soil fertility, intractable pests and diseases, climate change phenomena (drought and floods), social conflicts, and the lack of an enabling policy environment for agriculture.

To increase productivity, smallholder farmers who make up the bulk of the farming population in Africa need access to the best current technologies for crop improvement in conjunction with agro-input, such as seed delivery, efficient irrigation, appropriate mechanization, improved cultural practices, farm credit, and market support (Anthony and Ferroni 2012).

Modern biotechnology breakthroughs can be harnessed to address many of the intractable biotic (pests and diseases) and abiotic stresses (soil fertility declines) of modern agriculture.

Biotechnology has been defined by the Convention on Biological Diversity as "any technological application that uses biological systems or derivatives thereof, to make or modify products or processes for specific use".

Agricultural biotechnology encompasses the following activities:

- *Genomics, which includes among others, DNA characterization, gene discovery, and gene action in plants and their wild relatives as well as in animals.*
- *Molecular breeding or marker-assisted selection (MAS)* using genetic markers to rapidly speed up improvement and development of new, higher-yielding varieties that can tolerate drought, resist pests and diseases, and contain improved nutritional content. This approach can take years off the time it takes to breed new varieties and make them available to farmers. Application of MAS accelerates development of both conventional and GM crops.
- *Genetic modification or genetic engineering or transformation* to produce superior crops, animal products, vaccines, drugs, and other biologicals that address intractable problems that cannot be solved by other methods.

Genetic modification is the most powerful of the tools of modern biotechnology, and has been the most controversial. It involves the artificial transfer of genes from one living species of plant or animal to another within a given species or across the species barrier. It is the transfer across the species barrier that has caused concerns of safety to humans, animals, and to the environment. Products from such transfers may be designated as transgenic organisms, genetically modified organisms (GMOs), or genetically engineered (GE) organisms. The use of GMOs is governed by international protocols, notably the Cartagena Protocol on Biosafety (CPB), which covers the safety of transboundary movement of living modified organisms (LMOs). Many countries have national legislation to regulate the safe deployment of GM products. These vary in stringency from promotional legislation to very restrictive ones.

Genetically engineered products have had, by far, one of the most positive impacts on food security and the economies of countries that have adopted the technology over the 16 years of their commercial use.. Despite the lack of scientifically proven evidence of any adverse effects, many developing countries have not harnessed the technology to address their nations' intractable agricultural problems. An overwhelming number of African countries have not taken advantage of the technology to address the challenges to food production and poverty reduction. The few countries that have taken advantage of the technology in Africa, and elsewhere in the developing world, are deriving benefits from the technology, and may be considered as role models for those still skeptical about the deployment of the technology.

The major GM biotechnology adopters of the developing world include Argentina, Brazil, India, the Philippines, and South Africa.. The status of GM product deployment and its impact on their economies will be examined, as also the political will that led to the commercialization of GM products. There are useful lessons for other countries in Africa.

Given the challenges to food production alluded to above, there is the need to invest in agricultural technologies that will reverse the current low production practices fueled by the low investment in improved technologies. In many instances, production increases have been recorded through expansion of arable land rather than increased production from existing arable lands.

The new technologies involving GM approaches are used for the production of new varieties where genetic variation does not exist in the available crops. Conventional breeding methods alone cannot solve the intractable problems of African agriculture listed. Genetic transformation approaches coupled with marker-assisted breeding practices are needed to rapidly produce the new products required that enhance yield and nutritional quality. Increased investment in research and development activities that generate safe and high-quality plant and animal products from GM technology to complement conventional technologies will be needed to address the problems of food security and nutritional inadequacies in Africa.

Global Status of GM Crops

On a global scale, 160 M ha of GM crops were produced in 2011 (James 2011). Genetically modified crop production has seen a phenomenal 94-fold increase from 1.7 M ha in 1996 to 160 M ha in 2011. The James (2011) report indicates that 90 % of the 16.7 M farmers who planted GM crops were from developing countries. A total of 29 countries grew GM crops, 19 of these being developing countries. Over the years, the GM crops in commercial production have been herbicide-tolerant soybean, Bt cotton, Bt maize, and Bt canola. Two-trait stacked gene (BT/HT) varieties of these crops are in production in about a dozen countries.

The global area of GM crops is as indicated (Table 15.1). Countries growing in excess of 50,000 ha of GM crops are considered major cultivators (Table 15.2). They are designated as mega countries (James 2011).

Status of Biosafety in Africa

Only 12 countries in Africa have biosafety laws that allow the commercialization of GM crops. These are Burkina Faso, Ghana, Mali, Senegal, and Togo in West Africa, Kenya and Sudan in East Africa, Malawi, South Africa, Zambia, and Zimbabwe in southern Africa, and Egypt in northern Africa. Out of these countries, only Burkina Faso, Egypt, and South Africa have actually introduced commercial GM products such as Bt cotton in Burkina Faso and South Africa, and Bt maize in Egypt and South Africa. Sudan is pilot-testing Bt cotton production on a commercial scale preparatory to a launch. Numerous GM products are being developed for seven countries in Africa (Table 15.3).

Table 15.1 Global area of GM crops (million ha) in 2011

Crop	Hectares	Cropped area (%)
Soybean	75.4	47
Maize	51.0	32
Cotton	24.7	15
Canola	8.2	5
Sugar beet	0.5	<1
Alfalfa	0.2	<1
Papaya	<0.1	<1
Others	<0.1	<1

Source: James (2011)

Table 15.2 Countries growing in excess of 50,000 ha of GM crops in 2011

Country	Million ha (global %)	Commercialized GM crops
USA	69.0 M ha (43 %)	HT/Bt maize; HT soybean; HT canola; Bt/HT cotton VR squash; VR papaya; Bt/Ht potato; sugarbeet; HT alfalfa
Brazil	30.3 M ha (19 %)	HT soybean; Bt cotton; Bt maize
Argentina	23.7 M ha (15 %)	HT soybean; Bt/HT cotton; Bt /HT maize
India	10.6 M ha (7 %)	Bt cotton
Canada	10.4 M ha (7 %)	HT canola; HT/Bt maize; HT soybean; HT sugarbeet
China	3.9 M ha (2.4 %)	Bt cotton; Bt polar; PRSV papaya;VR sweet pepper; DR,VR tomato
Paraguay	2.8 M ha (2 %)	HT soybean
Pakistan	2.6 M ha (2 %)	Bt cotton
South Africa	2.3 M ha (1 %)	HT/Bt cotton; HT/Bt maize; HT soybean
Uruguay	1.3 M ha (1 %)	HT soybean; Bt maize
Bolivia	0.9 M ha (1 %)	HT soybean
Australia	0.7 M ha (<1 %)	Bt/Bt-HT cotton; HT canola; FC carnatiion
Philippines	0.6 M ha (<1 %)	Bt/HT maize
Myanmar	0.3 M ha (<1 %)	Bt cotton
Burkina Faso	0.3 M ha (<1 %)	Bt cotton
Mexico	0.2 M ha (<1 %)	Bt cotton; HT soybean
Spain	0.1 M ha (<1 %)	Bt maize
Total	160 M ha (100 %)	–

Source: James (2011)

In general, there is a parallel between GM product development and utilization, and the level of effective biosafety legislation. Effective legislation is one that is backed by the appropriate Legislative Instrument that assists the implementation of the law passed. Long drawn-out review processes for GM crop applications, even in the presence of the enabling legislation, have a negative influence on the access to GM technologies. The institutional arrangements to facilitate the implementation of the legislation must be in place. Political will must exist to create the enabling framework. Such an enabling framework exists in the seven African countries handling GM crops at the various stages of release (Table 15.3). By far the vast majority of African countries that also face food security challenges and could benefit from the engagement of modern biotechnology products and services do not have the appropriate legislation or the will to implement existing legislation.

Table 15.3 Status of GM crops in Africa in 2011

Country	Commercialized or trial release	Greenhouse trait	Field trial	Confined field trial
Burkina Faso	Cotton	NA	NA	Cowpea, sorghum
Egypt	Maize	Tomato	Cotton, wheat, potato	NA
Nigeria	NA	NA	NA	Cowpea, cassava, biofortified sorghum
Kenya	NA	Sorghum	Cotton	Maize, cotton, cassava, sweet potato
Uganda	NA	NA	NA	Cotton, banana, maize
South Africa	Maize, cotton, soybean	Cassava, sorghum	NA	Maize
Sudan	Cotton	NA	NA	NA

Source: Adapted from Conway (2012)
NA Not applicable

Lessons from Lead GM Biotechnology-Adopting Developing Countries

Countries such as Argentina, Brazil, India, and the Philippines have made significant strides in the development and use of modern biotechnology for socio-economic development. These, and several African countries, notably Burkina Faso, Egypt, and South Africa, have made advances in the harnessing of modern biotechnology for the advancement of their nations' agriculture, and have derived significant economic benefits from such engagement that, together with the lead countries listed above, serve as role models for the rest of Africa.

The steps taken by the lead biotechnology developing countries are captured vividly in the 2011 publication on the "Global Status of Commercialized Biotech/GM Crop: 2011" by Clive James of the International Institute for the Acquisition of Agri-biotech Applications (ISAAA).These countries cited above developed the political will to engage in the technology. The manner in which this was done will be examined to provide lessons for African countries that are slow to adopt modern biotechnology as an option in the advancement of agriculture for socio-economic development.

Brazil

Brazil, the most populous South American country, has a population of 194.2 million, and an arable land area of 59.6 M ha. The commercialized GM crop area is 30.3 M ha, with herbicide-tolerant soybean, Bt cotton, and Bt maize as the major GM crops (Table 15.2). It is the second largest GM crop producer in the world, with

20.6 M ha of the GM crop hectarage devoted to HT soybean. Genetically modified crops are a major contributor to the growth of the Brazilian economy. Over the 8-year period from 2003 to 2010, GM crops contributed US $4.6 billion to the economy of Brazil.

For Brazil and the other lead GM crop-producing countries, the political will started with the institution of an enabling regulatory framework that was implemented. In the particular case of Brazil, it experienced earlier difficulties in GM crops extending over 5 years. Such challenges are typical of many African country situations. For Brazil, the consolidation of the biotech regulatory framework and the effective function of the Brazilian National Technical Commission on Biosafety (CTNBIO) enabled Brazil to accelerate the approval of biotech events. In 2010 alone, Brazil approved a record number of eight products — six at the end of 2011. Thus, Brazil made up for the lost time in the first 5 years, and has currently approved 32 biotech traits for farm use — five for soybean, 17 for maize, nine for cotton and one for edible virus-resistant beans (James 2011).

A genetically modified Phaseolus bean resistant to Bean Golden Mosaic Virus developed by EMBRAPA (The Brazilian Agricultural Research Corporation) has been given approval for commercial release in 2011. The Golden Mosaic Virus is a devastating disease of bean. The GM *Phaseolus* bean was developed entirely by Brazilian scientists at EMPRAPA. This has been a developing country breakthrough.

Apart from the enabling biosafety framework and the will to implement it, the availability of highly trained manpower in plant breeding led to the development of many crops suited to various agro-ecologies. This presented a choice and thus ease of adoption to farmers.

Argentina

Argentina is the third largest producer of GM crops in the world (Table 15.2). Out of this, 19.1 M ha was soybean, with 3.9 M ha for GM maize and 0.7 M ha for Bt cotton. Argentina is the world's largest producer of soybean oil. Over the 1996–2010 period, farm income gained from GM crops was US $12.2 billion.

The entire 100 % of soybean grown in Argentina, 86 % of the maize, and 99 % of the cotton are genetically modified.

Argentina, like Brazil, has benefited from the accelerated approval of GM crops. Since the introduction of commercialized GM crops in 1996, the country has, as at 2011, approved the release of 21 biotech crop varieties. Four biotech crop varieties were released in 2011 alone.

The gross benefit accruing to the country from the adoption of GM crops over the 15-year period from 1996 to 2010 was US $72,363 million. HT soybean, the largest contributor, contributed US $65,153 M. Of the gross earnings, 72.3 % went to farmers, 21.3 % went to the government (from taxes collected), and 6.5 % to the technology providers (Table 15.4).

Table 15.4 Economic benefits of biotech crops (million US$) and percentage distribution in Argentina

		Amount (percentages) of benefits accrued to		
Crop and trait	Total benefits	Farmers	National government	Technology developers
HT soybean	65,153	47,105.0 (72.3 %)	13,877.6 (21.3 %)	4,169.8 (6.4 %)
Bt/HT corn	5,375	3,665.8 (68.2 %)	612.8 (11.4 %)	1,096.5 (20.4 %)
Bt/HT cotton	1,834	1,760.6 (96.0 %)	0	73.4 (4.0 %)

Source: James (2011)

In Argentina, 670,000 ha (98 %) of total cotton is genetically modified. Reasons for the increase in GM cotton over the last 5 years in Argentina are:

- Availability of better-adapted biotech varieties
- Increased returns
- Increased awareness among farmers, of the benefits of technology
- Improved reporting of events

Actions taken by Argentina over the past years have been a clear demonstration of political will to promote the adoption of GM technology. The actions that underscore Argentina's political will are:

- Early adoption

The country has been an early adopter of GM crops worldwide. For instance, herbicide- tolerant soybean was made available to farmers there at the same time it was introduced to the American farmer. An early adoption of GM crops is a strategic issue that calls for the early adoption of regulatory procedures as a pre-requisite. Early adoption gives the country a head-start.

- Availability of world-class plant breeders

This is required to develop crop varieties suited to various agro-ecologies, and ones that will give the farmer the chance to make a choice. It is the trained plant breeder that will introgress the new genes into elite varieties developed.

- Trained and innovative farmers

These farmers, exposed to new crop varieties including GM and agronomic procedures, would be more inclined to adopt new technologies in agriculture.

- Creation of a pioneer regulatory system.

Such a regulatory system led to the safe adoption of GM crops early on also accounted for the rapid growth in the economy from GM crop introduction.

India

India started Bt cotton production in 2002, with the cultivation of 50,000 ha. Bt cotton is now the most productive and profitable crop in India. In 2011, India cultivated 10.6 M ha of Bt cotton out of a total of 12.1 M ha cotton. Production is by 7 M farmers, with average farm size of 1.5 ha/farmer. The growth in total farm income for India over 1996–2010 was US $ 9.46 M.

Reasons for India's success with Bt cotton:

- Political will demonstrated by long-term planning in areas of capacity-strengthening for plant breeding and the adoption of appropriate biosafety legislation.
- India is the only country in the world where hybrid cotton is the principal commercial crop. India released its first commercial hybrid in 1970.
- The commercial approval of Bt cotton in 2002 was a breakthrough.
- Increased availability of long staple cotton varieties through breeding and selection. Long staple cotton is desired by the world market.
- Bt cotton is promoted as a multiple crop:
    - Cotton seed oil production for food
    - Cotton seed cake for livestock and poultry
    - Cotton fibre for the textile industry.

Following on from the success with Bt cotton, India is taking urgent steps to approve other GM tested crops such as Bt maize and eggplant (brinjal).

Political Statements by Leaders

Pronouncements by India's political leaders at the highest level of government underscore the political will to advance the course of promotion or otherwise of an advanced technology like modern biotechnology.

Thus, in July 2011, the Prime Minister Dr. Manmohan Singh called on agriculturists to judiciously use biotech to improve productivity and enhance farmers' income. He called for increased spending on agricultural research, increased irrigation facilities, and the promotion of biotech carefully to boost crop productivity and enhance farmers' incomes. The Prime Minister stressed the need for regulatory control on GM crops to be based strictly on scientific criteria (James 2011).

The Philippines

Currently, Bt maize is the only commercially released GM product in the Philippines. The area put under stacked Bt/HT maize in 2011 was 545,000 ha (James 2011).

Economic growth from Bt maize is estimated for 2010 as US $ 63 M. Bt maize was first commercialized in the Philippines in 2003.

Pipeline GM products nearing commercial release are:

- Golden rice
- Fruit and shoot borer resistant eggplant
- Papaya against the ring spot virus.

The Philippines is the only country in SE Asia to implement a regulatory system, and the first in Asia to approve and grow a GM major feed crop.

Concerns on biotech safety started in 1987 when DNA technology was considered one of the effective tools in research. University of the Philippines, Los Banos and International Rice Research Institute (IRRI) formed a joint committee on Biosafety Joint Committee, which formulated biosafety guidelines for R&D. The initiative was eventually adopted nationally by Executive Order No 430 series 1990 issued by the President. This established the National Committee on Biosafety in the Philippines (NCBP).

The demonstration of political will in the Philippines has been by the establishment of the NCBP by Executive Order. It shows the country's commitment to the use of biotechnology, identified by the Department of Science and Technology (DOST) as the flagship of leading edge technologies to be used as strategic tools to achieve sustained economic development. Subsequent Presidents after President Corazon Cojuangco Aquino continued to support biotechnology as a major focus of the country's R&D program (Gonzales et al. 2009).

On 16 July 2001, President Arroyo, in his biotech policy statement, stated in part, "We shall promote the safe and responsible use of modern biotech and its products as one of the several means to achieve and sustain food security, equitable access to health services, sustainable and safe environmental and industry development".

In 2001–2002, multi-location field tests at a time of national controversy over GM maize introduction unwittingly served as demonstration fields showing the efficacy of the gene transfer. The adage "seeing-is-believing" holds here.

South Africa

In South Africa, GMOs are regulated under the GMO Act 15 of 1997, and has been in operation since 1999. The Act is managed by the Ministry of Agriculture that appoints an Executive Council to review applications for the granting of release for field trials or commercial release. The Act has tried and tested provisions for handling disputes, as was the case in the granting of a permit for Bt maize in 2003. Allowing state institutions to work expeditiously in the handling of GMO applications or disputes is a mark of political commitment to the free workings of a regulatory process (Morris et al. 2005). South Africa could share its experience with biosafety regulatory processes with the rest of Africa, and could be so contacted on a need basis.

Currently, South Africa derives immense benefits from modern biotechnology. In 2011, the GM crop area was 2.3 M ha. Of this, 1.9 M ha or 72 % was for white maize, 450,000 ha was for Roundup-ready (RR) or herbicide-tolerant (HT) soybean, and 15,000 ha was under Bt cotton. Adoption rate for Bt cotton was 100 %, with 95 % of the cotton being stacked (Bt/HT). The estimated revenue from GM crops in 2011 for South Africa was US$133 M (James 2011).

Burkina Faso

Commercial Bt cotton production in Burkina Faso started in 2008, with the granting of the permit for its commercial release. In 2011, the total cotton area was 424,810 ha, 247,000 ha (58 %) of which was Bt cotton. A 5 % drop in the area of Bt cotton cultivated from the 2010 figure of 260,000 ha was noticed in 2011. This decline was due to increased fertilizer cost and conflicting, often wrong, extension advice and disagreement over pricing of cotton (James 2008). The following lessons were learnt:

- Need for favorable market prices
- Need for affordable inputs
- Adherence to good stewardship and appropriate agronomic practices

Currently, the average cotton holding per farmer in Burkina Faso is 3.25 ha. Net gain to the farmer from Bt cotton was $66 over the planting of conventional (non-Bt) cotton.

Burkina Faso has exhibited considerable will in the adoption of Bt cotton. The country was the first in West Africa to commercialize Bt cotton production in 2008.

In a statement to the National Peasant Federation in 2010, HE Blaise Campaore declared, among other things:

"In a continent that is hungry, the GM debate should be very different. The technology provides one of the best ways to substantially increase agricultural productivity and thus ensure food security to the people. ...But with falling cotton prices, we have no choice but to produce in quantity. And biotechnology may allow us to reach 2 to 3 million tons".

The challenge to Burkina Faso's political will to harness biotechnology will be put to test when products in the pipeline such as Bt cowpea and biofortified sorghum destined to enter the food chain request permits for commercial release."

Possible Reasons for the Slow Pace of GM Biotechnology Engagement in Africa

The single most important factor accounting for the slow pace of GM biotechnology engagement on the African continent is the lack of political will to advance the course of biotechnology. For all the lead biotechnology countries cited from the developing world (Argentina, Brazil, India, the Philippines, South Africa, and the up-and-coming Burkina Faso), the political commitment to biotechnology advancement was at the highest level of government, backed by positive statements by politicians from below the Head of State. These Heads of State emphasized commitment to biotechnology and the institution of science-based legislative processes, and general support for biotechnology capacity.

The commitment of Heads of State was translated into the support for enabling legal frameworks for biosafety. In all countries, there were challenges to the introduction of enabling legislation, but there was a steady move to get the legislation in place. Once the laws were in place, the lost time was made up for by rapid approvals following the review process for applications. For instance, Brazil released eight products in 2010 and six more in 2011. A rapid approval process also characterized the activities of the regulatory agencies in Argentina.

The early adoption of proven technologies like GM biotechnology at the farmer level would give the country a head start in a highly competitive industry. This accounted for the lead Argentina attained in Roundup-ready soybean production. Early government support would accelerate the pace of adoption of new technologies.

In the absence of specific legislation for biosafety, scientists in countries such as South Africa and the Philippines used existing legislation on pest and disease control to start confined trials. These actions were eventually taken up by newly introduced legislation, giving the scientists in those countries a head start.

Manpower development engaged the attention of the governments of the lead countries cited as mentors. Large numbers of well-trained plant breeders, in collaboration with molecular biologists, led to the release of large numbers of crop varieties adapted to various ecologies in the country, giving farmers a wide choice of plant varieties.

Awareness creation of biotechnology activities and transparency will facilitate adoption of technologies at the farmer level. For instance, the multi-locational trials on Bt maize in the Philippines created the “seeing-is-believing” effect on farmers, and led to rapid adoption of Bt maize in the country.

Way Forward for Africa

The adoption of GM technologies on a need basis is the way forward for Africa. Given the challenges from intractable pests and diseases, the need to adopt sustainable intensification processes in agriculture in view of expanding populations and dwindling arable land holdings and the threat of climate change, biotechnology offers powerful tools that can lead to the development of new safe products with genes coming from outside the existing traditional varieties. Modern biotechnology approaches will need to be harmonized with conventional plant breeding to develop the new crop varieties.

For GM biotechnologies to advance, the political will to move the technology, an enabling legislation with the needed institutions at work to fast-track new germplasm release, and general infrastructural support by way of agro-input delivery, extension delivery, and market support services should be in place. The lack of these support services has adversely affected the adoption of GM technologies in the respective countries.

Investment in the training of high level manpower in plant breeding, biotechnology, and biosafety will underpin the steps to advance in the course of biotechnology.

References

Anthony VM, Ferroni M (2012) Agricultural biotechnology and smallholder farmers in developing countries. Curr Opin Biotechnol 23:278–283

Conway G (2012) Africa in search of safe and high quality biotech crops. In: Africa in search of safe and high quality biotech crops. 1st Pan African on Stewardship of Agricultural Biotechnology. Forum for Agricultural Research in Africa (FARA), Accra

FAO (2012) Hunger statistics. www.fao/hunger/en

Gyase B (2011) NEPAD's initiatives for food security: the case for nutrition. Rural 21. Int J Rural Dev 45:27–29

Gonzales LA, Javier EQ, Ramirez D, Carino FA, Baria AR (2009) Modern biotechnology and agriculture: a hisotry of the commercialization of biotech maize in the Philippines. STRIVE Foundation, Manila

James C (2008) Global status of commercialised biotech crops: 2008. Brief No. 39. ISAAA, Ithaca

James C (2011) Global status of commercialized biotech/GM crops: 2011. ISAAA Brief No. 43. ISAAA, Ithaca

Morris EJ, Van Rensburg JBJ, Hoffman JH, Lazarus P (2005) Good regulatory practices for developing countries—lessons learned in South Africa from an appeal under the GMO Act. Paper presented at the 9th ICABR international conference on agricultural biotechnology: ten years later. 6–10 July 2005, Ravello

Chapter 16
Influencing Politicians and Policy Makers for a Viable Biotechnology Sector: A case Study of the Nigerian Biosafety Bill Drafting, and Passage Process at the Parliament

Bamidele Ogbe Solomon, Rufus Ebegba, and Rose Suniso Maxwell Gidado

Abstract As worldwide adoption of genetically modified crops (GMOs) continues to grow at a double-digit rate, there remains uncertainty, confusion, apprehension, and resistance to the technology in some African countries including Nigeria. Noteworthy, however, is that the Federal Government of Nigeria adopted biotechnology policy in April 2001 and later in November of the same year; the National Biotechnology Development Agency was formed as a parastatal under the Federal Ministry of Science and Technology. The Agency was set up to domesticate and promote modern Biotechnology tools and products. The safe practice of this technology requires a biosafety law in place. The Cartagena Protocol on Biosafety requires parties to it to develop their Biosafety Administrative and Regulatory Framework in order to effectively regulate activities of agricultural biotechnology. A National Biosafety Framework under the UNEP/GEP by the Federal Government of Nigeria, which included, among other things, the National Biosafety Policy Biosafety Bill, was developed in 2006. The House of Representatives and Senate passed the Draft Biosafety Bill in July 2010 and June 2011 respectively. This Bill got transmitted to President Goodluck Ebele Jonathan for assent but was not assented to because the tenure of the National Assembly that passed it ended a day after the Bill got passed. It has been returned to the 7th Assembly, which is in Session right now and receiving attention after which President Goodluck Ebele Jonathan will finally pass it. This paper attempts to present the experience and the road map of the enactment of Biosafety Law in Nigeria.

B.O. Solomon (✉)
Director General's Office, National Biotechnology Development Agency (NABDA), Abuja FCT, Nigeria
e-mail: bosconsult77@yahoo.com

R. Ebegba
Biosafety Office, Federal Ministry of Environment, Abuja FCT, Nigeria

R.S.M. Gidado
Open Forum on Agricultural Biotechnology and Biotechnology(OFABBA) Unit, National Biotechnology development Agency(NABDA), Abuja FCT, Nigeria

F. Wambugu and D. Kamanga (eds.), *Biotechnology in Africa*, Science Policy Reports 7, DOI 10.1007/978-3-319-04001-1_16,

Keywords Agricultural Biotechnology • Biosafety • Commercialization • Genetically modified Organisms

Abbreviations

AATF	African Agricultural Technology Foundation
ABNE	Africa Biosafety Network of Expertise
ARCN	Agricultural Research Council of Nigeria
BAG	Biosafety Advocacy Group (Nigeria)
BIC	Biotechnology Information Centre (Nigeria)
FMARD	Federal Ministry of Agriculture and Rural Development
FME	Ministry of Environment (Nigeria)
FMENV	Federal Ministry of Environment
GMO	Genetically modified organism
IFPRI	International Food Policy Research Institute
IITA	International Institute of Tropical Agriculture
LMO	Living modified organism
NABAC	Nigeria Agriculture and Biotechnology Project Advisory Committee
NABDA	National Biotechnology Development Agency (Nigeria)
NABP	Nigerian Agricultural Biotechnology Project
NBC	National Biosafety Committee (Nigeria)
NBF	National Biosafety Framework (Nigeria)
NRCRI	National Root Crops Research Institute (Nigeria)
OFAB	Open Forum on Agricultural Biotechnology in Africa
PBS	Program for Biosafety Systems
REFORMS	Restructured Economic Framework for Openness, Reform, and Macroeconomic Stability
S&T	Science & Technology, Ministry of (Nigeria)
UNEP/GEF	United Nations Environment Programme/Global Environment Facility
USAID	United States Agency for International Development
USDA	US Department of Agriculture

Introduction

Modern biotechnology has been identified as an important tool that can help nations to achieve food sufficiency/food security, industrial growth, health improvement, and environmental sustainability, which will ultimately lead to economic growth, and job and wealth creation. All these attributes will have a positive impact on sound economy, enhanced stable polity, and social harmony under a biosafety legal framework. The focus of this paper is on the lessons learnt on the Passage of Nigerian Biosafety Bill into an Act by the Nigerian Parliament.

The Nigeria National Biosafety Bill provides a legal framework for the deployment and domestication of modern biotechnology in Nigeria. This includes the provision for establishment of a National Biosafety Management Agency to manage the emerging Act. Prior to the passage of the Biosafety Bill, all line federal ministries, agencies and departments, and stakeholders made inputs into it, and adopted it. The Biosafety Act, which is the outcome of the Bill, will therefore provide a legal impetus for the regulation of modern biotechnology activities. The adoption of modern biotechnology under a legal framework would assist research institutes and other organizations that have the mandate to promote biotechnology activities in the fulfillment of their set mandates by using the technology to enhance food security, fight malnutrition, alleviate poverty and hunger, and enhance the environment.

Having a biosafety law is the only sure guard against indiscriminate dumping of GMOs into the environment and market, to ensure that products of genetically modified organisms (GMOs) are certified fit for the environment and human health.

The experience of the creation of a Biosafety Act in Nigeria dates back to 1994 when the first National Biosafety guidelines were produced as interim regulatory instruments in Nigeria. This paved way for the development of a National Biosafety Framework in 2001; this consists mainly of Biosafety Policy and the Biosafety Draft Bill.

(a) Biosafety Guidelines:

Nigeria started biosafety activities with the development of the following guidelines:

(i) *The 1994 Biosafety Guidelines*: these Guidelines were developed by the Federal Ministry of Agriculture, and were specifically for the agricultural sector pre-Cartagena Protocol on Biosafety. They were, however, not put into use because they did not meet the required Nigeria biosafety standard.
(ii) *The 2001 National Biosafety Guidelines*: these Guidelines were developed through the participation of relevant stakeholders, involving line government agencies, non-governmental organizations (NGOs), and individuals knowledgeable in biodiversity conservation, biotechnology, and biosafety. It was developed after the signing of the Cartagena Protocol on Biosafety, and it was in line with the protocol.
(iii) *National Biosafety Framework (NBF)*: in line with the Protocol (signed in 2000 and ratified by Nigeria in 2003), which requires countries that are parties to it to develop national administrative and legal frameworks as a way of domesticating the Protocol, Nigeria developed a National Biosafety Framework in 2006, which consists of, among others, the Biosafety Bill and the Biosafety Policy. The NBF is to ensure the safety of the practice of modern biotechnology, handling, and use of its products (GMOs) that may have adverse effects on conservation and sustainable use of biodiversity, taking into account risks to human health.

(iv) *National Biosafety Policy (2006)*: The overall objective of the Policy is to provide a regulatory regime and guidance for the sustainable development of the science of modern biotechnology, its application, and safe use of its products without prejudice and risk to public health, environmental health, national sovereignty, human dignity, and fundamental human rights.

Stages Involved in the Drafting of the Biosafety Bill

The National Biosafety Legislation basically sought to develop a holistic mechanism for the safe research, application, and use of modern biotechnology aimed at identifying and minimizing the potential adverse effect on the conservation and sustainable use of biological diversity, including risk to human health.

The drafting of the Biosafety Bill passed through various stages. Some of the stages included the following:

(i) *Review of relevant international treaties*: this process involved the review of the international conventions and protocols, of which the Convention on Biological Diversity and the Cartagena Protocol on Biosafety were considered to be the major ones. This is in recognition of the interdependence of the different environmental components, their impact on the ecosystem, and the fact that "environment" has no national boundaries. The First Earth Summit in Rio de Janeiro, Brazil, adopted a global approach as the best strategy for the conservation, management, and protection of the environment. The resolutions adopted at the Summit resulted in the development of several international conventions and protocols on environment.

(ii) *Review of existing relevant national laws*: This process entailed reviewing existing national legislation, regulations/guidelines that impact on the use of modern biotechnology including research, development, safety, movement, and general commerce in GMOs and living modified organisms (LMOs). The existing domestic legal frameworks were evaluated under two broad categories—administrative and legislative. The third category would have been judicial pronouncement, but case law on biotechnology/biosafety in Nigeria is virtually non-existent. Nigeria has several laws and regulations on environment in general, but these laws only tangentially touch on the use of biotechnology. The National Policy on Environment 1989, which was reviewed in 1999, and the Nigeria Biosafety Guidelines 2001, were the only instruments that had some bearing on biotechnology/biosafety. However, some of the legal/administrative instruments on agriculture, industry, trade, occupational health, and safety are acknowledged to have some relevance on biotechnology and biosafety.

Development of National Biosafety Framework. In 2002, a National Committee was inaugurated by the Federal Government of Nigeria under the supervision of the Federal Ministry of Environment to develop a National Biosafety Framework for the country. The Committee was made up of

16 members drawn from line federal ministries, civil society groups, and individuals with cognate experience on issues of biotechnology and biodiversity conservation. In carrying out its duties, the committee had various stakeholders' consultative meetings and workshops. A Draft Biosafety Bill and Biosafety Policy were two of the documents that the committees came up with. These documents were further subjected to various stakeholders' review sessions. The Draft Biosafety Bill was in particular subjected to a broad range of legal experts
nationally and internationally. The committee finally submitted the documents to the government in 2006.

(iii) *The Nigerian Agricultural Biotechnology Project*: A Memorandum of Understanding was signed in May 2004 between the Governments of Nigeria and the USA, as well as the IITA, to initiate a program called the Nigerian Agricultural Biotechnology Project (NABP). The project had been in progress since September 2003, and was made up of three major components:

i. Improved research and development on crops (cowpea and cassava) and livestock
ii. Improved biosafety implementation and field testing of GM crops
iii. Improved awareness and public acceptance

The National Biotechnology Development Agency (NABDA) was the secretariat for the project, and directly involved in the awareness component of the project. A sub-agreement was signed with IITA, which was the direct implementation partner of the project. A Biotechnology Information Centre (BIC), which was situated at NABDA, was established under this agreement. The BIC was designed to provide a facility for accessing biotechnology information. It was also proposed to develop BICs in the Six Zonal Centers. The NABDA also handled the hosting of the Nigeria Agriculture and Biotechnology Project Advisory Committee (NABAC), which was set up to provide the necessary guidance and direction for the successful implementation of the project. These two activities were being funded by the project.

Awareness Workshops: These were planned for the six Geopolitical Zones. All the zones were covered during the Agreement period as well as the extension of the awareness workshop, which was made in order to reach more people. Other awareness workshops held included those for Directorate Cadre Civil Servants in the Ministries of Science & Technology (S&T), the Federal Ministry of Agriculture and Rural Development (FMARD), and the Federal Ministry of Environment (FMENV).

The Draft Biosafety Bill was developed under this program, as it provided funding for experts from the USA to assist and guide the drafting committee. Under the same program, funds were also provided for the premier application for a confined field trial in Nigeria. This was for a transgenic cassava that was transformed at the Danforth Plant Science Laboratory to confer resistance to the African Cassava Mosaic Virus. The application was filed jointly by NABDA and

IITA, Ibadan, with the test site in an IITA field approved by the Ministry of Environment. Unfortunately, the project had to be truncated before actual field planting when a reversal occurred, which was adduced to the gene gun method employed for the transformation. The group in the Danforth Laboratories made a fresh attempt using the more efficient *Agrobacterium tummafacien* transformation method. After this initial failed attempt, several successful trials involving the three crops—cassava, corn and cowpea—have been carried out using the 2001 Biosafety Guidelines.

The Parliament Experience, How It All Started

The Restructured Economic Framework for Openness, Reform, and Macroeconomic Stability (REFORMS) Project funded by USAID, Nigeria, which handled the Biotechnology Policy Program in Nigeria, organized a Sensitization Workshop in November 2008, which was attended by various stakeholders, institutions, and individuals. The aim of the workshop was to highlight the potential of agricultural biotechnology as a tool for ensuring food security and wealth creation in Nigeria.

The workshop resulted in a communiqué, which noted that:

- Agricultural biotechnology has great potential to ensure food security and wealth creation by enhancing crop yield, improving the nutritious value of food, and providing a cleaner environment.
- Biotechnology is still at the infancy stage in the country, despite Nigeria's acceptance and commitment towards exploiting its potential.
- The absence of a Biosafety Law was hampering the implementation of the biotechnology policy of the country, and it would be good to have the bill passed into law soon.

The workshop also observed that:

- The intended bill benefited from the experience of two earlier documents.
- Biosafety guidelines 1994 and 2001 were approved by the Federal Executive Council, but the intended bill was yet to be presented to the Federal Executive Council by the Ministry of Environment.

The Communique

- A Biosafety Advocacy Group (BAG), whose membership should include major stakeholders, was set up to work towards the passing of the Biosafety bill.
- Public enlightenment sensitization and awareness campaigns should be intensified and targeted at the relevant audience.

- The membership of the Biosafety Committee should be carefully re-constituted to include major stakeholders such as the National Agricultural Seed Council, scientists from universities and research institutes, private sector organizations, and farmers' organizations.
- Regulations and strict guidelines in vital areas such as confined field trial, commercial release, laboratory practices, labeling, inspection, and GMO waste disposal should be developed.
- The Ministry of Environment should be more proactive, and embark on advocacy and sensitization programs on biosafety for both the Federal and State Legislatures.
- The cost of processing applications should be USD 5,000 for applicants within Nigeria, and USD 12,000 for applicants from outside the country, and applications that will be submitted within the first 5 years after passing the Bill should be co-funded by the Federal Government.
- An online Biosafety Clearing House should be established and made accessible to the public, with some level of confidentiality on trade-related secrets.
- Communication lines must be immediately opened to small farmers and agro businesses to bring sufficient awareness and education on all aspects of biotechnology and their benefits, its perceived drawbacks, what the government is doing, what the other countries are doing, what scientists and different stakeholders are doing, and what the global business and the multilateral organizations are doing.
- Guaranteed Minimum Price Policy of government as it relates to food prices should be assured.
- The Intellectual Property bill before the National Assembly should be passed into law as soon as possible to facilitate the implementation of the biotechnology policy of the government.

Presentation of the Bill to the House of Representatives

It was based on the recommendations in the Communiqué above, that a meeting with the then Chairman, House of Representative Committee on Agriculture, Hon Peter Gbenga Makanjuola, was arranged. This meeting had two officials from each of the following: Ministry of Environment (FME), The National Biosafety Committee (NBC), NABDA, and the United States Agency for International Development (USAID) REFORMS Project. This meeting was made possible by the Special Assistant to the Chairman, Mr Sunday Owolabi, who attended the workshop in Jos, Nigeria. The members of NABDA and the Federal Ministry of Environment explained to the Chairman what biosafety was all about, what it stood for, its importance to the implementation and development of biotechnology to Nigeria, and how it had been passed in other African countries.

This led to the resolve of the Chairman of the House Committee on Agriculture to pick up the Bill as a privately sponsored Bill in January 2009, after several unsuccessful attempts in earlier years, to present the Bill to the Parliament as an Executive Bill. He predicated his action on the need to support research aimed at developing modern biotechnology that will enhance farming systems aimed at optimum food production with land resource protection, and which are compatible with the socio-economic conditions of Nigeria.

Consequently, the Biosafety Bill went through the first reading in the House of Representatives, after which it was committed to the House Committees on Agriculture, Environment, Science, and Technology for review so as to make recommendations for the consideration of the plenary. As part of its parliamentary processes, a public hearing on the bill was organized, where different stakeholders attended and provided input into the bill.

The Bill went through the first and second readings as well as public hearing, after which a retreat was convened by the Program for Biosafety Systems (PBS) and NABDA for collation of reports and memoranda, and finally, on 20 July 2010, the House of Representatives passed the bill and moved it to the Senate.

At the Senate, the Bill was referred to the Senate Committees on Agriculture, Science, and Technology. Initially, the Bill received a major setback at the Senate as the Chairman Committee on Environment and Ecology opposed it, stating that she had a similar Bill awaiting consideration at the Senate. Moving forward was made possible through serious Net mapping and networking to involve the former President, Honourable Olusegun Obasanjo, former Minister of Science and Technology Professor Turner T Isoun, Senator Smart Adeyemi who became a major spokesman, the Senate President, David Mark, and other well-disposed and highly connected citizens, before being finally passed at the Senate on 1 June 2011, which was practically the last seating of that Senate for political dispensation.

The passage of the bill at the Senate brought joy and relief to the Biosafety Advocacy Group (BAG) Team, as a failure to pass a bill could have sent out a very bad message.

Communication

The communication strategy for agricultural biotechnology in Nigeria is the institution of the program tagged Open Forum on Agricultural Biotechnology in Africa (OFAB), an initiative of the African Agricultural Technology Foundation (AATF), Nairobi, Kenya. The Open Forum is a platform to facilitate the flow of information from the scientific community to policy makers and the general public. It brings together stakeholders in biotechnology/biosafety—scientists, journalists, the civil society, farmers, industrialists, law makers, and policy makers—and enables interactions among them. The Open Forum takes the form of a monthly lunch meeting that provides an opportunity for key stakeholders to know one another, share knowledge and experiences, make new contacts, and explore new avenues of bringing the benefits of biotechnology to the African agricultural sector.

The objectives of OFAB are to:

1. Establish and manage a range of platforms that will raise the profile of biotechnology usage in agriculture for enhanced agricultural productivity.
2. Contribute to informing policy decision-making processes on matters of agricultural biotechnology through provision of factual, well-researched, and scientific information, to be able to participate effectively in policy discussions.
3. Forge strategic alliances to create synergy and optimization of resources through convening and encouraging inter-institutional networking and knowledge-sharing in the agricultural biotechnology space.
4. Enhance targeted capacity-strengthening that will improve communication across all sectors interested in biotechnology for African agricultural development.

The first OFAB was launched in Nairobi, Kenya, on 14 September 2006. This was followed by the Uganda Chapter in December 2007, Tanzania Chapter in May 2009, Egypt Chapter in August 2009, and Ghana Chapter in August 2011. In Nigeria, NABDA, which has the mandate to promote biotechnology activities that positively respond to national aspiration and food security, job/wealth creation, affordable healthcare delivery, and sustainable environment, is the host organization for OFAB, while the Agricultural Research Council of Nigeria (ARCN) is the co-host.

Since the inception of OFAB in Nigeria it has been hosted in Abuja, with the aim of sensitizing legislators, key government officials (policy makers), civil society, media, scientists, and other stakeholders to the importance of domesticating modern biotechnology, and to the need for a Biosafety Law in Nigeria. It has contributed immensely to mitigating the negative public perception about the domestication of this technology. The Open Forum stirred the passage of the Biosafety Bill at the National Assembly in 2010 and 2011 respectively. Its sensitization program enlightened the legislators and the general public on the need for a Biosafety Law, despite the negative influence of environmentalists and poor public understanding of biotech operations and procedures. With the Bill ready for assent, the regulatory framework once in operation will ensure the commercialization and approval of biotech products.

The hosting of OFAB in Abuja has recorded over 100 participants on a monthly basis, making up to 1,000 participants a year. Three special sessions of OFAB were held at the National Assembly, as well a Special session of Media OFAB and Poster OFAB in 2010/2011. All the sessions included top government officials—ministers, members of House of Representatives, senators, permanent secretaries, directors-general/CEOs, executive directors, executive secretaries, and directors. The first OFAB hosted outside Abuja (in Umudike) in collaboration with the Program for Biosafety Systems (PBS) and the National Root Crops Research Institute (NRCRI) recorded an attendance of over 300 participants. This was basically due to an increased number of stakeholders (researchers and farmers) located outside the Federal Capital Territory.

This advocacy campaign is much needed, and considering its success so far in the Federal Capital Territory (Abuja), the hosting of OFAB outside Abuja became imperative in order to enlighten a wider audience. Nigeria has a population of over 160 million people, and the Federal Capital Territory (Abuja) with a population of just over 5 million people is not a good representation of the Nigerian people as most research institutes, industries, and local farmers—who are the major players in the biotechnology industry—are located outside Abuja.

As a consequence, OFAB, Nigeria through NABDA commenced the hosting of the OFAB events in the six geopolitical zones of Nigeria. These are the North-Central, North-West, North-East, South-South, South-West, and South-East regions. This will cover the period of 3 years from 2012.

So far, three sessions have been held in three geopolitical zones of the country—Ibadan (Oyo State), Owerri (Imo state), and Makurdi, (Benue State). It also held Special Sessions recently inside the Capital City: (i) for the Seed Sub-sector, to sensitize and enlighten stakeholders in the seed industry on the eminent adoption of agricultural biotechnology and the use of genetically enhanced seeds and seedlings to increase agricultural productivity in the country, and (ii) for the media, which provided an opportunity for interaction between the media and scientists on effective communication and reporting on GM-related issues.

Why OFAB Is an Effective Means to Addressing This Need

The Open Forum is well-placed to contribute towards supporting and mitigating negative public and political perceptions regarding biotechnology, especially GMOs. It is independent of other organizations, programs, and projects, and is unencumbered by institutional bureaucracy. Although an initiative of AATF, it is operated at the country level through local institutions that provide it with the much-needed local view and appreciation of key issues. Discussions provide the potential for generating home-grown solutions towards encouraging biotechnology acceptance. The Open Forum's operating model allows for deep country contextual knowledge, which potentially gives it much credibility with local stakeholders, more than an outside organization would. The Open Forum is also multi-stakeholder and multi-disciplinary, drawing on a wide range of expertise. It also enjoys significant convening power, demonstrated through its track record of organizing regular monthly meetings, which attract large/diverse audiences and prominent speakers. These factors mean that OFAB is flexible, responsive, impartial, inclusive, and authoritative.

Information-Sharing Platform

Nigeria uses communication outlets provided by the new media so as to capture a wide spectrum of the country. These include:

- Facebook
  - The OFAB Nigeria Facebook page helps network and reach people from all over the world; it helps create awareness, and serves as an advertising medium for the upcoming events. Address on Facebook: Open Forum on Agricultural Biotechnology
- Twitter
  - In order to reach wider and younger audience, OFAB Nigeria created a twitter account. The address is twitter: @OFAB Nigeria
- Newsletters
- Radio jingles
- Advocacy visits to media houses
- Newspapers and magazines
- Television broadcasts
- One-on-one engagement with policy makers, farmers, media, civil societies, etc.

Below are some of the faces at some of the various OFAB sessions in Nigeria.

From L-R: Dr. Kenneth Nwachukwu, the then Executive Director, NCRI and Prof. B.O. Solomon, former DG/CEO, NABDA at the Umudike OFAB in Abia State

Middle: Former Minister of Agriculture and Rural Development, Prof. Abubakar Sheikh and Other Dignitaries at the Umudike OFAB in Abia State

Second from *Left*: The former DG/CEO, RMRDC, the former DG/CEO, CPC, the former DG/CEO, NABDA; also second from *Right*: Mr. Russ Nicely, USDA all in Group Photograph after the Consumer Protection Council (CPC) OFAB

From *Left*: Prof. B.Y. Abubakar, Executive Secretary, ARCN, Abuja, Nigeria and Prof. B.O. Solomon and Sharon Pauling of USAID, Abuja at the CPC OFAB

Group Photograph at the December 2010 Senate OFAB

Middle: Senator Hambagda sharing thoughts at the December 2009 Senate OFAB Session

Prof. B.O. Solomon and Prof Diran Makinde sharing knowledge during the February 2011 Senate OFAB Session

A cross Section of Dignitaries at February 2011 Senate OFAB

Ancillary Activities Towards the Passage of the Biosafety Bill

To build the capacity of the legislators so as to enable them to effectively address issues related to biosafety and modern biotechnology, the following activities were executed:

(a) Study tours to countries involved in the commercialization of GM crops were organized for the Chairmen of the House Committees that the Bill was referred to. Some of the countries visited included Burkina Faso, The Philippines, South Africa, and USA.
(b) Sensitization workshops were organized for members of the various committees that the Bill was referred to.
(c) Monthly OFAB was organized to showcase and sensitize legislators, as well as the general public, on the benefits and potentials of biotechnology to national development and the need to have a Biosafety Law for Nigeria.

Facts about the Passage of the Biosafety Bill at the Parliament

Some facts on the passage of the Biosafety Bill include:

(a) The drafting of the Bill took 4 years, from 2002 to 2006 after which it was subjected to Review Sessions for an additional year taking it to 2007.
(b) The Bill was later presented to the then Chairman, House Committee on Agriculture, Honourable Peter Gbenga Makanjuola as a Private Bill in 2009.
(c) The Parliamentarians did not initially understand the biosafety issues surrounding the Bill.
(d) Although the process of drafting of the Biosafety Bill was begun by the Federal Ministry of Environment, the National Biotechnology Development Agency championed the process of its passage by the Parliament, working closely with other major stakeholders in the National Biosafety Office of the Federal Ministry of Environment.
(e) On arrival of the bill after its passage at the House of Representatives to the Senate, the sponsor of the Bill at the House of Representatives still saw the need for him to continue leadership in the facilitation of the Bill, which was not regarded as ideal from the point of view of some of the senators, particularly the Senate Committee Chairperson on Environment and Ecology. At this point, she felt the Bill was an Environment Bill and as such, that she should take the lead.
(f) One other major setback at the Senate was that there were two other bills on Biosafety being sponsored by two different Senators, one by the Chairperson of the Environment and Ecology Committee, and the other by the Vice Chairman of same Committee. She felt her Bill should be the one to be passed by the Senate, and not the one from the House of Representatives. This became worrisome to the major stakeholders who felt the one from the House of

Representative was more holistic, since it had passed through stakeholders' review and the cross-section of the House of Representatives who took time to debate the Bill. The Bill generated serious debate at the second reading in the Senate; it was, however, referred to the Committees on Agriculture and Science and Technology for detailed review before presentation to the Plenary. The Chairperson, Environment and Ecology openly opposed the Biosafety Bill on the Senate floor during the second reading, but the Bill finally went through.

(g) The final passage of the bill took place at the eleventh hour to the end of a Parliamentary tenure on 1 June 2011; this was another challenge, but was overcome. This was not without serious lobbying. If the Senate had failed to pass the Bill, this would have meant starting the process of the passing of the Biosafety Bill afresh. The bill was passed on the understanding that it would lead to an agricultural boom, and also generate employment in Nigeria if modern biotechnology is practised under a legal framework. The view was also taken that the research Institutes in Nigeria will take advantage of a Biosafety Law to be more innovative.

(h) Some environmental non-governmental organizations held various public sessions to discredit the Biosafety Bill, and called for a moratorium on the bill. Some sections of the public remained confused on the debate on the bill, and some remained passive and indifferent, while the proponents of modern biotechnology remained in the forefront of the passage of the Bill.

Major Stakeholders in the Process, Leading to the Passage of the Biosafety Bill

1. Federal Ministry of Environment: piloted the drafting of the Biosafety Bill with the participation of the following Federal Government Ministries and Agencies: Federal Ministries of Agriculture, Education, Foreign Affairs, Trade and Investment, Nigerian Customs Service, Nigeria Food and Drug Administration and Control Agency, Nigeria Institute for Veterinary Research.
2. National Biotechnology Development Agency: championed the process of the passage of the Bill, by providing funding and overall direction of the activities that led to the passage of the bill.
3. USAID: through workshops and sensitization programs and study tours
4. Program for Biosafety Systems/IFPRI: through funding of sensitization workshops and programs.
5. AATF: through funding study tours for Parliamentarians and Biosafety Regulators from the National Biosafety Office. Also through funding OFAB.
6. Agricultural Research Council of Nigeria: through co-funding of OFAB
7. Institute for Agricultural Research: co-funding sensitization activities
8. National Root Crops Research Institute: co-funding sensitization activities

9. Donald Danforth Plant Science Center: by funding study tours for Parliamentarians and Biosafety Regulators from the National Biosafety Office.
10. Africa Biosafety Network of Expertise (ABNE): by providing funds for sensitization program for Nigerian Parliamentarians, and training workshop for National Biosafety Committee Members and Biosafety Regulators in general
11. The Director General of the National Biotechnology Development Agency, Prof Bamidele Solomon, was highly instrumental in the passage of the Bill through his sterling leadership.
12. Mr. MPO Dore, Consultant Programme for Biosafety Systems (PBS); and Mr. Awoniyi James Olatunji, Member of the Agricultural Transformation Agenda (ATA)'s Cassava Value Chain Group, Federal Ministry of Agriculture and Rural Development.
13. Mrs. Rose Gidado and other staff of NABDA by providing administrative and technical support
14. Staff of National Biosafety Office for professional and technical support

Status of the Biosafety Act

Upon getting the bill passed by the National Assemblies, it was transmitted to the Legal Unit at the Presidency, from where it was forwarded to the Ministries of Science and Technology and Agriculture and Rural Development. Both Ministries gave very timely and positive responses stating the benefits to the country.

In the light of this, the Biotechnology Awareness Program has been intensified by reaching out to the people beginning with the nation's capital, Abuja, and going to the various geopolitical zones of the country with additional support from AATF and USDA. The responses have demonstrated that there is great expectation from rural dwellers who are seeking for solace from their unrewarding toil characterized by minimal return on their investment, both in cash and in kind on the subsistence farms.

In preparation for the Passage of the Bill by the present Assembly and consequent assent by President Goodluck Ebele Jonathan, the following activities are on-going:

- Drafting of Biosafety Guidelines: the Programme for Biosafety Systems, the Federal Ministry of Environment, and the National Biotechnology Development Agency organized a workshop to draft the biosafety guidelines.
- Scientists, the Media, Farmers, Civil Servants, Public Servants, Lawyers, Regulators, Custom, Immigration etc
- Advocacy and Awareness: the Nigeria Chapter of OFAB has expanded its advocacy programs by hosting the forum across the country's (geopolitical zones to promote biotechnology and biosafety in Nigeria)

Conclusion

When the present Assembly passes the Bill again, the next stage will be for President Goodluck Ebele Jonathan to give his assent. Nigeria cannot afford to discard this Bill, as it will deprive the country of the great potential modern biotechnology has to offer in the fields of agriculture, medicine, industrial development, and environmental sustainability.

Chapter 17
Conclusion and Way Forward

Florence Wambugu and Daniel Kamanga

Abstract The depth and breadth of the papers contained in this book confirms that Africa has made a critical shift from general discussion and focus on the safety of plant biotechnology to a concrete exploration of crops and traits deemed useful for various African countries. Overall, Africa continues to make good progress towards biotech acceptance. There are many scientifically promising projects underway using genetically modified crop technologies to improve African crops, mostly by public research institutions. While these projects often generate state of the art results in the laboratory, they invariable suffer setbacks when the R&D moves from the contained laboratory, greenhouse and contained field trial studies into the highly regulated biosafety and regulatory stage due to limited funds and often expertise, to drive product development and commercialization.

Keywords Research and development • Genetically modified organisms

Abbreviations

Bt	*Bacillus thuringiensis*
GM	Genetically modified
R&D	Research and Development

The path to commercialization is littered with many challenges, among them: There is limited or lack of in-house regulatory expertise to work in the highly regulated environment that governs GM crops. Many public–private-partnership projects are naïvely optimistic about how regulations will be applied to their public good

F. Wambugu (✉) • D. Kamanga
Africa Harvest Biotech Foundation International (AHBFI), P.O. Box 642,
Nairobi 00621, Kenya
e-mail: fwambugu@africaharvest.org; dkamanga@africaharvest.org

F. Wambugu and D. Kamanga (eds.), *Biotechnology in Africa*,
Science Policy Reports 7, DOI 10.1007/978-3-319-04001-1_17,

projects. Funding and capacity challenges mean that public institutions usually work only on one project, thereby limiting the institutional learning that private technology developers have. Very costly biosafety and regulatory dossier development for deregulation and commercialization of the crop or product. The decentralised nature of public research institutions works against the development of a multidisciplinary critical mass in the field of regulatory affairs, stewardship, and product acceptance.

Once the early results of GM work leave the contained environment of laboratory and glasshouse, they become much more visible, because they require detailed scrutiny from regulatory bodies, and their intentions are heavily scrutinised. They also start being targeted by professional, well-resourced, well-informed anti-biotechnology activists. This comes as a shock to many projects, many of whom walk into this regulatory and public acceptance arena unprepared on how both the authorities and the opinion makers will treat their good intentions to deliver products of biotechnology for public good.

There are lessons that can be learnt from the relative success of private sector biotechnology developers. These lessons show that, without serious investment, the support of a critical mass in regulatory affairs, government affairs expertise for getting political goodwill, excellent issues management strategy to manage the anti-GM lobby activism, and a well-resourced public outreach strategy, projects are likely to fail in their primary objective of delivering public or private good, biotechnology crops and products. Indeed, to date nearly all biotechnology-derived crops and products in the market come from the multinational private sector for these reasons. The success story of Bt cotton in Burkina Faso clearly shows the need and the value of political will, combined with local regulatory and public support, that within a relatively short time can bring the benefits of private sector biotechnology to the small-scale farmers of Africa.

There is need to continue supporting continued cultivation and trade in South Africa, Burkina Faso, Egypt, and Sudan—the four countries where biotech crops are commercialized in Africa to date. This is not an easy endeavor, given on-going challenges, especially with regard to an unpredictable regulatory landscape. For example, in South Africa, (at the time of writing this section) contentious labeling issues are likely to be part of the amendments to the GMO Act. Burkina Faso managed to stave off a push for regulatory changes that would have included unfriendly liability clauses, due to strong political will.

In our view, African governments must provide the required political leadership. Also, both multinational companies and the African seed sector, which stand to benefit most from biotech adoption, must be more committed to Africa. The African public research sector, farmers, and consumers are the pillars on which success will rest. Unfortunately, their trust equity has not been adequately harnessed. Instead, disharmony exists, even within those organizations that support the technology. Funding agencies should ensure that international and Africa-based organizations are supported in an equitable way that enhances harmony. There is urgent need to support a wider base of African local organizations to demonstrate the benefits of biotechnology and to build local expertise of African scientists.

As a way forward, we believe that a biotech product development centre of excellence, with local and international expertise, whose sole mandate is to provide expert support for all African biotech projects, in critical areas such as biosafety and regulation, would bring economics of scale and be an important game-changer. Its aim would be to assist a wide range of public goods projects by supplying a common source of expertise in those parts of the project which are common to all projects, and for which most public institutions do not have the internal capacity or experience. We are aware that some African organizations have been funded to play some of these roles; however, the current lack of synchrony does not augur well for the continent. The paradox is that the success of one of the projects could be undermined by the failure of another.

The proposed approach would help African institutions to apply the kind of learning experience which provides large private sector companies with their competitive advantage in bringing products of biotechnology to the market. It would allow African research institutions to focus on their strengths while providing the required support to the product development centre as needed, to come up with target products The biotechnology product development centre would concentrate resources and expertise to assist all relevant projects on challenges such as biosafety assessment and regulatory file management, regulatory compliance management, regulatory policy support, communication and issue management, socio-economic benefit analysis, Government affairs and such.

Along with these pockets of challenge are positive stories. Ghana in 2011 developed a Biosafety Law, and is using approved Biosafety data from Burkina Faso to pave the way for commercialization of Bt cotton. Malawi, Uganda, and Zimbabwe have great prospects of commercialization of GM crops. As the above diagram shows, Africa has the potential to increase the number of commercialized countries from the current four to nine in the next 3 years. This will only happen if more countries adopt science- and risk-based biosafety and regulation, while building public confidence in the regulatory systems and with unwavering political support for crops and products that will benefit the economy of Africa.

FSC
www.fsc.org
MIX
Papier aus verantwortungsvollen Quellen
Paper from responsible sources
FSC® C105338

If you have any concerns about our products, you can contact us on
ProductSafety@springernature.com

In case Publisher is established outside the EU, the EU authorized representative is:
Springer Nature Customer Service Center GmbH
Europaplatz 3, 69115 Heidelberg, Germany

Printed by Libri Plureos GmbH
in Hamburg, Germany